RECONSTRUCTING
CITY POLITICS

Cities & Planning Series

The Cities & Planning Series is designed to provide essential information and skills to students and practitioners involved in planning and public policy. We hope the series will encourage dialogue among professionals and academics on key urban planning and policy issues. Topics to be explored in the series may include growth management, economic development, housing, budgeting and finance for planners, environmental planning, GIS, small-town planning, community development, and community design.

Series Editors

Roger Caves, Graduate City Planning Program,
  San Diego State University

Robert Waste, Department of Political Science,
  California State University at Chico

Margaret Wilder, Department of Geography and Planning,
  State University of New York at Albany

Advisory Board of Editors

Edward J. Blakely, *University of Southern California*
Robin Boyle, *Wayne State University*
Linda Dalton, *California Polytechnic State University, San Luis Obispo*
George Galster, *Wayne State University*
Eugene Grigsby, *Univeristy of California, Los Angeles*
W. Dennis Keating, *Cleveland State University*
Norman Krumholz, *Cleveland State University*
John Landis, *University of California, Berkeley*
Gary Pivo, *University of Washington*
Daniel Rich, *University of Delaware*
Catherine Ross, *Georgia Institute of Technology*

David L. Imbroscio

RECONSTRUCTING CITY POLITICS

Alternative Economic Development
and Urban Regimes

Cities & Planning

SAGE Publications
International Educational and Professional Publisher
Thousand Oaks   London   New Delhi

For information address:

SAGE Publications, Inc.
2455 Teller Road
Thousand Oaks, California 91320
E-mail: order@sagepub.com

SAGE Publications Ltd.
6 Bonhill Street
London EC2A 4PU
United Kingdom

SAGE Publications India Pvt. Ltd.
M-32 Market
Greater Kailash I
New Delhi 110 048 India

Printed in the United States of America

Library of Congress Cataloging-in-Publication Data

Imbroscio, David L.
    Reconstructing city politics: alternative economic development
and urban regimes / David L. Imbroscio.
        p. cm   (Cities and planning; v. 1)
    Includes bibliographical references and index.
    ISBN 0-7619-0612-6 (acid-free paper). — ISBN 0-7619-0613-4 (pbk.:
acid-free paper)
    1. Urban policy—United States. 2. United States—Economic
policy. 3. City planning—United States. 4. Municipal government—
United States. I. Title. II. Series.
HT123.I43 1997
307.76'0973—dc20   96-35645

97 98 99 00 01 02 03 10 9 8 7 6 5 4 3 2 1

Acquiring Editor:	Catherine Rossbach
Editorial Assistant:	Nancy Hale
Production Editor:	Michèle Lingre
Production Assistant:	Karen Wiley
Typesetter/Designer:	Janelle LeMaster
Design Director:	Ravi Balasuriya
Print Buyer:	Anna Chin

To Rebecca

CONTENTS

PART III: CONCLUSIONS

F o r e w o r d

The study of cities is a dynamic, multifaceted area of inquiry which combines a number of disciplines, perspectives, time periods, and numerous actors. Urbanists alternate between examining one issue through the eyes of a single discipline and looking at the same issue through the lens of a number of disciplines to arrive at a holistic view of cities and urban issues. The books in this series look at cities from a multidisciplinary perspective, affording students and practitioners a better understanding of the multiplicity of issues facing planning and cities, and of emerging policies and techniques aimed at addressing those issues. The series focuses on both traditional planning topics such as economic development, management and control of growth, and geographic information systems. It also includes broader treatments of conceptual issues embedded in urban policy and planning theory.

The impetus for the *Cities and Planning* series originates in our reaction to a common recurring event—the ritual selection of course textbooks. Although we all routinely select textbooks for our classes, many of us are never completely satisfied with the offerings. Our dissatisfaction stems from the fact that most books are written for either an academic or practitioner audience. Moreover, on occasion, it appears

as if this gap continues to widen. We wanted to develop a multidisciplinary series of manuscripts that would bridge the gap between academia and professional practice. The books are designed to provide
valuable information to students/instructors and to practitioners by
going beyond the narrow confines of traditional disciplinary boundaries to offer new insights into the urban field.

David Imbroscio's *Reconstructing City Politics: Alternative Economic
Development and Urban Regimes* represents the inaugural book in this
series. We are excited about this book and feel it offers an important
bridge between two distinct literatures—economic development and
regime theory. It "grounds" the urban regime theory discussion in
"praxis" considerations such as: Where do I go from here both to
understand *and* to change cities? He challenges the parochialism of both
left and right-wing contemporary planning advocates by arguing si-
multaneous for urban growth and development—with the accompanying panoply of theme malls, ballparks, and convention centers—
and the need to insure a parallel commitment and expenditure to better
plan for the children, schools, safety, jobs, and life opportunities of the
poorer residents of America's center cities. He argues that progressive
planning can and must be done if cities are to plan successfully for the
challenges ahead in the 21st Century. Imbroscio develops an interesting
line of literature and research in "alternative economic develoment
policy." In doing so, his book significantly advances our understanding
of this vital subject, providing insights that are equally valuable to
planning/policy scholars, students, and practitioners. We believe this
well-written and well-documented book will generate a vigorous new
discourse on economic development that challenges both theory and
practice.

—Roger W. Caves,
San Diego State University

—Robert J. Waste,
California State University at Chico

—Margaret Wilder,
State University of New York at Albany

PREFACE

This book springs from the idea that the study of central city politics has advanced sufficiently to provide us with a good understanding of the key empirical processes and dynamics at work in the urban polity. Therefore, it strives to move the bounds of research forward by beginning an investigation of two central issues that, with the existing knowledge in hand, now can be fruitfully addressed. The first of these issues involves the normative implications arising from these empirical processes and dynamics; the second involves the possibilities for altering these processes and dynamics—that is, for a "reconstruction" of central city politics. Otherwise put, this book asks, simply, what is wrong with the workings of contemporary urban politics, and how might it be fixed?

Proceeding in this intellectual fashion, this project has been inspired in general form and structure by a body of scholarship concerned with matters beyond the city. This work is composed of studies that (a) provide a careful diagnosis of normative problems arising from the empirical workings of current political-economic institutions and, in light of this analysis, (b) attempt to develop reconstituted institutional arrangements to correct for these normative problems. Among these works of "institutional design," I have found the writings of Robert Dahl (1985), Stephen Elkin (1987), Benjamin Barber (1984), and John Dryzek (1987) to be particularly inspiring.

More specifically, I attempt to make a unique and valuable contribution to the study of the city by bringing together and building on two strands in the urban literature that have developed largely in isolation. The first body of literature is composed of the fine work conducted over the past two decades by a group of scholars concerned with the systematic analysis of how the interplay between political and economic factors shapes urban life. This work in "urban political economy" includes the writings of my teachers Stephen Elkin (1985) and Clarence Stone (1989), and those of Susan Clarke (1987), Scott Cummings (1988), Richard DeLeon (1992a), Susan Fainstein (Fainstein et al., 1983), Barbara Ferman (1996), Roger Friedland (1983), Mark Gottdiener (1987), Richard Hill (1983), Bryan Jones (Jones & Bachelor, 1986), Dennis Judd (1984), Paul Kantor (1988), John Logan and Harvey Molotch (Logan & Molotch, 1987), John Mollenkopf (1983), Paul Peterson (1981), Adolph Reed (1988a), Martin Shefter (1985), Gregory Squires (1989), Todd Swanstrom (1985), and Ronald Vogel (1992).

The second body of literature, generally less developed, is composed of the work of scholars who have thought seriously about *localized* alternatives for building (what is deemed to be) a more just and democratic urban economy. In various ways, this concern has been informed by the writings of Gar Alperovitz (1990), Murray Bookchin (1982), Severyn Bruyn (1987), Pierre Clavel (1986), Peter Eisinger (1988), Gerald Frug (1980), Judith Garber (1990), Robert Giloth (1988), Edward Goetz (1990), Christopher Gunn (Gunn & Gunn, 1991), Norman Krumholz (1991), Staughton Lynd (1987a), James Meehan (1987), Robert Mier (1993), David Morris (1982a), Andrew Polsky (1988), Derek Shearer (1989), and Wim Wiewel (Wiewel & Weintraub, 1990).

In this project I refer to these localized efforts as "alternative" urban economic development strategies. Such strategies include a broad array of economic development initiatives differing in significant ways from the approaches now commonly employed in cities, for example, so-called "corporate center" (see, for example, Hill, 1983; Levine, 1988; Robinson, 1989) or "mainstream" (see Clavel & Kleniewski, 1990; Nickel, 1995) economic development strategies. As we will see, alternative strategies have a different set of actors and goals, and they potentially have a different effect on the structure of the urban political economy.

The argument of the book unfolds as follows: I begin by demonstrating that a consensus exists in the urban political economy field regarding the empirical nature of contemporary central city politics (or "urban regimes") in the United States. This consensus revolves around two widely observed tenets. First, an alliance between two sets of actors—key local public officials and those business interests heavily tied to the city—often manages urban governance. Second, these actors usually orient the city's policy agenda strongly toward the goal of achieving local economic growth via a set of traditional development strategies. I next discuss the normative implications of this consensus. I find that these two key tenets of city politics (or urban regimes) exacerbate political inequality among the urban citizenry and this, in turn, damages the health of our liberal democracy. I close the first chapter by arguing that the rectification of these normative problems is largely a matter of altering or "reconstituting" the nature of urban regimes.

I begin Chapter 2 by pointing out that if we are to understand how such a reconstitution can be brought about, we must understand why urban regimes form as they currently do. I then trace current urban regime formation to two broad structural features of the urban polity: the "external economic dependence" and "internal resource dependence" of city public officials. The first refers to the need for city officials to attract and retain mobile economic investment; the second, the need for those officials to garner extrastate resources from the local community in order to govern their cities effectively. Reconstituting urban regimes—and hence increasing political equality in the city—entails a lessening of these two structural dependencies. I conclude the chapter by showing that the reorientation of city economic development strategies in certain alternative directions has the potential to reduce these dependencies.

The remainder of the book explores that potential. After providing an overview of the vision and substance of three such alternative strategies—what I call entrepreneurial mercantilism, community-based economic development, and municipal enterprise—Chapters 3 through 5 assess their effectiveness and feasibility. I review a wealth of evidence suggesting their possible effectiveness. But I also find, using case studies, that formidable barriers hamper each strategy's feasibility. In the final chapter, Chapter 6, I explore whether these barriers can be overcome and, more generally, the likelihood that the strategies can bring about the requisite empirical and normative reconstruction of central city politics.

The guiding premise of this project is that the successful reconstruction of central city politics cannot be based on the promotion of explicitly antigrowth urban policies. Too often those urbanists seeking to achieve fuller democratic control and social justice in the city—goals consistent with the normative aspirations of this study—fervently endorse such an agenda. In contrast, I argue that unless reconstruction efforts can establish the necessary economic foundation, they are bound to fail. Development strategies must be reoriented, but economic growth must remain a central concern. In his brief but penetrating review of Stephen Elkin's (1987) work, Raphael J. Sonenshein (1988, p. 37) captures the essence of this perspective: "Elkin suggests that the opponents [of Paul Peterson's 1981 progrowth argument] are more interested in opposing capital investment than in finding a way to make the system work." This position, Sonenshein correctly notes, challenges both the Left and the Right in current urban political analysis.

More urban scholars, even those with a strong anticapitalist orientation, are beginning to appreciate the need for an economic base to underpin efforts at political reconstruction. The "flaw" in progressive analysis, writes the urban social theorist Susan Fainstein (1990) "is that it does not offer a [progressive] formula for growth" (p. 41). The political empowerment of social democrats in most cities—aside from places such as Santa Monica or Toronto, which have to fend off private capital —is unlikely to succeed without such a formula, she says (Fainstein, 1990, p. 43). Likewise, after an exhaustive study of "urban populism" in Cleveland, Todd Swanstrom (1985, p. 244) expressed similar sentiments: "The goal," he suggests, "should not be to eliminate growth politics but to subject it to the will of the majority."

Finally, I hope that this work, beyond its more scholarly and theoretical contributions, also will advance a more practical political purpose. I believe that the collected materials and data on alternative economic development strategies can serve as a guidepost and reference source for progressive urban regimes. Typically, these regimes come to power without a coherent vision of alternative economic development. They often know what they oppose—but not what to do in its stead. Below,

I describe an array of alternative approaches and policies, and I demonstrate how and why they should be pursued. For those progressive regimes inclined to embark on such a course, this work may, in some fashion, help them better find their way.

—David L. Imbroscio
Louisville, Kentucky

ACKNOWLEDGMENTS

The journey undertaken to complete this project has been long and, at times, arduous. Reaching the final destination could not have been accomplished without the encouragement, assistance, support, and inspiration provided by several people.

This book grows out of my doctoral dissertation at the University of Maryland—College Park. As the dissertation manuscript took shape, numerous colleagues and friends read and commented insightfully on various parts of it. I especially thank Judy Garber, Jeff Henig, Jyl Josephson, Manabi Majumdar, Jeff Spinner, and Kieron Swaine. Jeff Spinner also undertook the heroic task of reading and commenting on the entire dissertation, offering numerous helpful suggestions for its revision into a book. Furthermore, he provided friendship and encouragement during the many years in which I worked on this project. In this regard, I also wish to again thank Kieron Swaine. Throughout graduate school and beyond, Kieron's friendship and support were important sources of personal and intellectual sustenance.

The following people provided various unpublished materials, without which this book could not have been completed: Pierre Clavel, Barbara Ferman, Marion Orr, Jeff Shavelson, Larry Soderholm, Todd

Swanstrom, and Thad Williamson. The field research was aided greatly by the helpful and supportive efforts of Moe Coleman, David Morris, Alberta Sbragia, Mark Vander Schaaf, Gary Shiffman, and Marian Yee. I also thank the dozens of people I interviewed for taking time out of their busy schedules to talk with me about various aspects of this project.

Funding has been provided by the Department of Government and Politics at the University of Maryland, and I thank the department's chair, Jon Wilkenfeld, for these resources. Additional funding came from the Conley H. Dillon award, a prize given to the best dissertation being prepared in American Politics at the University of Maryland. For the final stages of this project, generous financial assistance was provided by the President's Research Initiative program and the College of Arts of Sciences at the University of Louisville.

I am further indebted to the many scholars who were mentors to me when I was in graduate school. John Dryzek, who at the time, was on the faculty at Ohio State, was an important early influence. During the developmental phase of this project, Gar Alperovitz, of the National Center for Economic Alternatives, provided some much-needed guidance and encouragement. Karol Soltan, Ronald Terchek, and Bill Hanna, all members of my dissertation committee at Maryland, each offered helpful suggestions on how the manuscript could be improved.

Special thanks go to Professors Clarence Stone and Stephen Elkin. Clarence has aided me in countless ways. Most important for this project, he has helped me to think more clearly about the nature of contemporary urban politics, something he knows more about than perhaps anyone else in the country. To Steve, my adviser at Maryland, goes my deepest gratitude for mentorship. Steve's intellectual imagination and his command of the fields of urban politics, political economy, and institutional design were a crucial source of guidance and inspiration that I freely drew on to conceive and develop the key ideas and themes of this book.

This project has proved to be rewarding in numerous ways. One of the most important rewards was the critical commentary, suggestions, and encouragement I received from many of the same scholars whose contributions influenced and shaped my thinking as a graduate student. As a young scholar, I felt fortunate and honored that such distinguished senior colleagues were willing to take the time to read and comment extensively on my manuscript. In this regard, I thank Pierre Clavel, Scott Cummings, Chris Gunn, Dennis Judd, Laura Reese, Todd Swanstrom, and Hal Wolman.

I thank Scott Cummings also for recommending the manuscript to Sage Publications. Likewise, I am grateful to Roger Caves, Margaret Wilder, and, especially, Bob Waste for warmly embracing this project for their new book series. Many thanks also go to Catherine Rossbach, my editor at Sage, who could not have been more helpful and supportive.

For the completion of this book, I was aided profoundly by the supportive academic environment provided by the Department of Political Science at the University of Louisville. This environment works to foster and sustain the research efforts of the department's faculty, especially its younger faculty. I particularly wish to thank Susan Matarese, Rodger Payne, Chuck Ziegler, and Paul Weber for their encouragement, advice, and support. Paul's support, as department chair, proved especially helpful, as he allowed me release time from teaching duties to finish this book. In addition, the department's graduate assistant, Rich Puszczewicz, furnished the technical support necessary to create many of the graphics found in the book, and Gayle Collins provided research assistance. Many thanks go to both of them. Above all, I wish to highlight my debt to one member of the political science department—Ron Vogel. Ron has been a constant source of advice and encouragement, and a much appreciated and valued combination of friend and colleague.

I wish to acknowledge the support provided by my family. My academic accomplishments are attributable directly to my father's efforts to instill in me from an early age the value of education. For these efforts, I am forever grateful. In a sense, the genesis of this project can be traced to the many discussions around the dinner table we had while I was growing up. These discussions most often focused on the lack of economic justice in our society and how such justice might be achieved. I am also greatly appreciative of the support provided by my brother, Michael. In light of his own academic and professional accomplishments, earning his respect has been an important source of encouragement.

Finally, and most deeply, I wish to thank Rebecca Dernberger. Rebecca read much of the manuscript in its various forms, and she provided diligent editing and helpful suggestions on clarity and style, as well as substance. Even more crucially, this project benefited immensely in less tangible ways from the much-needed emotional support and comfort she constantly provided over the years. For all of this, and more, it is to her that I dedicate this book.

PART I

Background

EMPIRICAL AND NORMATIVE FOUNDATIONS

Following in the tradition of Alexis de Tocqueville and John Stuart Mill, a normative understanding of the practice of city politics—and local political life more generally—should not be isolated from an understanding of the larger political order of which it is part (Elkin, 1987, pp. 1-4). In particular, the practice of local politics has a particular role to play in buttressing the foundations of this order. The mode of normative analysis presented here deems this larger political order in the United States—that is, "liberal democracy"—to be a desirable political way of life. Thus, the key analytical question becomes, What supports (or works to maintain) such an order (cf. Ceaser, 1990); and more specifically, how must local politics in the United States be practiced so that it contributes to this goal (Elkin, 1987)?

In this chapter, I argue that one key manifestation of local political life—political practice in central cities—is marked by substantial political inequality among its citizens, and that this hampers the flourishing of the larger liberal-democratic order. Hence, the practice of central city

3

politics must be "reconstructed" so that it better supports what is normatively desired—that is, a liberal-democratic political way of life.

URBAN POLITICAL ECONOMY
AND THE EMPIRICAL CONSENSUS

Two decades of urban research from a political economy perspective have produced, in broad outline, a compelling empirical account of the nature of governance in American central cities.[1] Although intellectual disagreements—paradigmatic, methodological, and ideological— persist, a consensus among scholars has evolved regarding the basic features of central city politics (see Vogel, 1992, pp. 12-21). Capturing the thrust of these studies in urban political economy, Elkin (1985) summarizes that "any enduring pattern in city politics . . . will revolve around an alliance between city politicians and the array of businessmen concerned with land use in the city" (p. 25). This alliance, he adds in a later work, will in turn come together around and be dedicated to a political agenda oriented toward intensifying the city's land use via the pursuit of local economic growth (Elkin, 1987, pp. 36-40). Though the precise nature of these enduring political patterns (or "urban regimes") varies according to time (Elkin, 1985) and the complexities of place (Cummings, 1988; Stone & Sanders, 1987), this growth-oriented, business-state governing alliance is a central feature of local political life in urban America.[2]

The nature and extent of this consensus is quickly discerned from a perusal of the two seminal, most influential theoretical statements in the burgeoning field of urban political economy: Harvey Molotch's (1976) article "The City as a Growth Machine" and Paul Peterson's (1981) book *City Limits*. Although radically different in their normative orientations and ultimate conclusions, each paints a similar picture of contemporary central city politics and policy: In essence, both argue that (a) the important matters of urban politics are decided largely by a locally based, public-private elite and (b) much of urban public policy is an exercise in employing growth strategies designed to buttress a city's economic fortunes.

Thus, as the urban political economy research teaches us, business interests play a key role in the governance of central cities. Of particular importance is one segment of the local business community: those economic enterprises heavily tied to and thus dependent on the "political economy of place."[3] The relative immobility of capital investment gives the controllers of these enterprises a strong incentive to be deeply concerned about, and hence participate in, the political life of the city (Friedland & Palmer, 1984; see also Cox & Mair, 1988; Humphrey, Erickson, & Ottensmeyer, 1989; Logan & Molotch, 1987). Though not usually in direct control of the local state, these "land-based business interests," as they might be called, have appeared as especially attractive allies to urban public officials. As a result, in city after city the pattern has been for urban public officials to gravitate toward these interests and form strong political alliances with them.

As for the second element of this consensus—the general pervasiveness of growth-oriented policies in central city politics—the level of interest in economic development matters among city officials over the past few years is best described as an "obsession" (Brintnall, 1989). As Eisinger notes in a study of the issue, no other policy area has commanded the kind of sustained attention in subnational politics, achieving a "high, often unique, place" on political agendas (Eisinger, 1988, pp. 19-20; see also Bingham & Blair, 1984).[4] One of the primary achievements of the political economy "turn" in the study of city politics is the explicit recognition that public officials' concern with local business performance is crucial to any adequate understanding of that politics.

Moreover, growth strategies empirically pervasive in central cities tend to take predictable forms. Variously labeled as "corporate-center" (Elkin, 1987, pp. 96-97; Hill, 1983, pp. 102-106; Levine, 1988, pp. 118-120; Reed, 1988a) or simply "mainstream" (Clavel & Kleniewski, 1990, pp. 203-208; Nickel, 1995, p. 373), these currently employed growth strategies emphasize (a) bricks-and-mortar (i.e., physical) approaches to economic development, usually calling for large-scale changes in land use patterns in or near the central business district, and (b) the public provision of financial incentives to attract investment into the city. More generally, we can trace the philosophical roots of these corporate-center/mainstream strategies to what Barnekov and Rich (1989, p. 213), following Warner (1968), call the "cultural tradition of privatism—a tradition that historically has tied the fortunes of cities to the vitality of their private sectors and encouraged a reliance on private institutions [rather than on public control and planning] for urban development" (cf. Leitner & Garner, 1993, pp. 59-60; Robinson, 1989, p. 285; Squires, 1994, pp. 91-96).

This scholarly consensus, as defined by these two key tenets, provides the empirical outline of contemporary central city politics in the United States. As I shall demonstrate below, however, the consensus exists at a descriptive level only: In the realm of explanation, urban political economists advance widely contrasting accounts of *why* urban politics looks this way. The dimensions of that dispute will be explored in the next chapter.

THE URBAN REGIME CONCEPT

The urban political economy literature conceptualizes the relatively stable political patterns manifest in central city politics as "urban regimes." Though this abstraction has been used as an analytical tool by various scholars in contrasting ways (see DiGaetano, 1989; Elkin, 1985; Fainstein et al., 1983; Stone, 1989; Swanstrom, 1988), in this study an urban regime refers to (a) the public officials and private interests that function together as allies in the city's *governing coalition*[5] and (b) the nature of the policy agenda pursued by this coalition (cf. Stone, 1989, pp. 3-9). This specification of the regime concept encompasses the two key premises of urban regime analysis: that urban governance must be understood in broad terms to include both state actors and actors situated in the larger urban community, and that the regime is, as Stone (1989) notes, "purposive, created and maintained as a way of facilitating action" (p. 4). Specifying the concept in this manner thus captures both the *who* and the *what* dimensions of urban politics: who exercises political power in the local decision-making process and what those with power attempt to accomplish.

The consensus sketched above regarding the current empirical state of contemporary central city politics points to the existence of what I shall call the *dominant regime form* in urban politics. This dominant regime form, then, is constituted by (a) a local governing coalition having at its center a close alliance between officials of the local state and land-based businesspeople,[6] and (b) an urban agenda heavily favoring corporate-center/mainstream growth strategies.

THE NORMATIVE IMPLICATIONS
OF THE EMPIRICAL CONSENSUS:
POLITICAL INEQUALITY IN CENTRAL CITIES

Over the past few decades, urban political analysis—unlike much contemporary social science research—has, quite heroically, demonstrated a robust affinity for normative matters. Such matters were, at root, of central importance in the earlier studies of community power and remain central in the more recent urban political economy research (Vogel, 1992). It is surprising, then, that so few studies make these normative concerns explicit and subject them to direct, consistent, and rigorous analysis. Otherwise put, it is surprising that most urbanists—albeit preoccupied with normative questions—actually eschew political theory (but see notable exceptions such as Dahl, 1967; Elkin, 1987; Garber, 1990; Long, 1962, 1986; Monti, 1990).

The existence of substantial political inequality, my focus in this study, is not the only normatively salient feature of central city politics. For example, in a similar analysis Garber (1990) cites the absence of "distributive justice" in the contemporary central city. Nevertheless, political equality has a special role to play in any polity committed to popular government. The eminent political scientist Robert Dahl (1985, p. 5) explains that political equality, though not the only important value in a good society

> is surely one of the most crucial, not only as a means of self-protection but also as a necessary condition for many other important values, including one of the most fundamental of all human freedoms: the freedom to help determine, in cooperation with others, the laws and rules that one must obey.

Indeed, though urban scholars often fail to investigate the concept systematically, a wealth of literature strongly implies that citizens in central cities are politically unequal. Representative of this literature is Shefter's (1985) excellent historical study of fiscal crises in New York City: Democratic government in major cities, he argues, is largely unrealized—as the structural context in which urban political practice occurs severely constrains the "democratic impulse" (p. 235). Likewise, the findings of numerous other major studies of central city politics broadly concur with this normative conclusion (see Kantor, 1988; Logan & Molotch, 1987; Stone, 1989; Swanstrom, 1985).

The Idea of Political Equality

To explore the thesis that a fundamental inequality afflicts urban politics, we must first specify with precision the nature of the basic concept. In its common theoretical conception, the idea of political equality necessitates that democratic institutions afford citizens equal power to influence political outcomes (Beitz, 1989, p. 4). Whenever citizens differ in their abilities to influence political outcomes, a condition of political *in*equality among citizens results.[7]

This formulation, though familiar, is inappropriate for the analysis provided here. Most crucially, it exhibits extreme insensitivity to the notion that the appeal of political equality stems not from an abstract and isolated justification of the concept's inherent normative attractiveness, but rather from its role in promoting a larger political way of life deemed desirable (cf. Ceaser, 1990, p. 95). That political way of life, as already suggested, is liberal democracy. And liberal democracy—as with any larger political order—encompasses a complex set of values and institutional relationships that must be maintained for it to flourish (see Elkin, 1993).

The problem with the common view, then, is that the kind of wholesale political reconstruction necessary to establish a political equality based on absolute equal power among citizens would no doubt violate some crucial values and institutional relationships of a liberal-democratic order: Most fundamentally, such a society, although perhaps democratic, would almost certainly not be liberal. Concretely, the autonomy of the family (cf. Fishkin, 1983) and the right to own and control private property (cf. Gutmann, 1980), among other institutional freedoms, would likely have to be sacrificed to provide the necessary egalitarian social and economic conditions ensuring that every citizen was truly as powerful as all others.

Yet any polity aspiring to be democratic requires some version of political equality (Dahl, 1982, 1985). The trick is to develop an understanding of the concept that also preserves liberal values.

Michael Walzer (1983, 1984) provides one compelling solution to this puzzle. Walzer suggests a "complex" version of equality. The key to this "complex equality" is "the art of separation"—ensuring that the various institutional settings or "spheres" in society (political, economic, social, religious, etc.) remain autonomous (Walzer, 1984, p. 320). He writes that complex equality "means that no citizen's standing in one sphere . . . can be undercut by his standing in some other sphere." (Walzer, 1983, p. 19). Though inequalities among citizens *within* each institutional setting may exist, inequality in this complex sense stems only from those circumstances where "separations don't hold"—when success (or failure) in one institutional setting or sphere is *convertible* into success (or failure) in other such settings (Walzer, 1984, p. 321).

From this general formulation of equality one can derive a consistent notion of equality in political life, what Walzer (1983, p. 310) calls "complex equality in the sphere of politics." Specifically, political equality—as understood here—is based not on the notion of equal political power among citizens, but on political institutions' freedom from encroachments emanating from society's other spheres.[8] These encroachments—following the logic of separation put forth by Walzer—occur when, for example, wielders of economic, social, or religious power gain a disproportional share of influence in the affairs of the state by virtue of their standing in these nonpolitical spheres.[9]

This version of political equality better corresponds with the theory and practice of liberalism. Most simply, it does not demand that all or most inequalities in society be somehow eradicated, and therefore poses significantly less of a threat to individual freedom. Furthermore, the project of separation—the construction of "walls"—itself can be understood as the essence of liberalism: "Liberalism is a world of walls," Walzer (1984, p. 315) argues, "and each one creates a new liberty." For example, as Walzer (1984, p. 315) illustrates, liberating the church from state control and vice versa—that is, erecting a "wall" between church and state—fosters both religious and political freedom. "Under the aegis of the art of separation," Walzer (1984) adds, "liberty and equality go together" (p. 321). Hence violations of the autonomy of the political sphere are as much cases of political unfreedom as they are cases of political inequality. Successful separation, however, eliminates both—uniting equality and liberty, and giving birth to a genuinely liberal conception of egalitarianism in political life.

The Empirical Consensus and
Political Inequality in Cities

How, then, does the consensus regarding the current empirical state of American urban politics point to the existence of a substantial degree of political inequality in central cities? As specified above, this consensus implies that two key features constitute the *dominant urban regime* form: a governing alliance between local public officials and land-based business interests, following an agenda marked by the aggressive pursuit of corporate-center/mainstream policies for growth. In light of this empirical reality, two separate but interrelated problems for local political equality emerge. These problems closely parallel the two tenets of the dominant urban regime form (i.e., the "who" and "what" dimensions of central city politics).

The first problem occurs because certain citizens—namely, land-based businesspeople—occupy a special position in urban regimes. I refer to this outcome as *the problem of privileged voices*. The second problem —what I call *the problem of economic inequality*—stems from the nature of the agenda pursued by the local governing alliance. Specifically, the regressive distributional effects of the current set of public policies adopted by central cities tend to reinforce and extend an extreme level of material deprivation among certain citizens, which is ultimately incompatible with a condition of political equality. I explicate each of these problems below.

The Problem of Privileged Voices

The problem of privileged voices arises from the strong attraction between local officials and land-based business interests in the city's governing coalition (cf. Stone, 1989). As noted earlier, the reasons *why* public actors in central cities gravitate toward local land-based interests are a matter of intense dispute; this will be the subject of further inquiry in the next chapter. Yet broad agreement exists about the basic existence of the familiar governing alliance between these two sets of actors. Although this alliance varies in strength from city to city (see Stone & Sanders, 1987)—and in selected places or extraordinary but brief times may be wholly absent (see Clavel, 1986; Rosdil, 1991)—the general pattern can be characterized as systemic (Elkin, 1985).

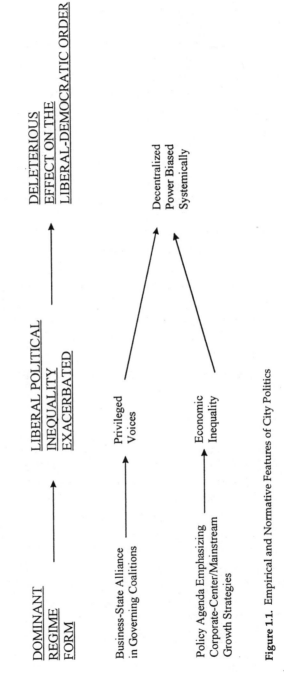

Figure 1.1. Empirical and Normative Features of City Politics

11

Two additional points regarding the nature of this alliance need to be made. First, the alliance is an arrangement between two independent sets of actors with distinct interests. Tensions can and do develop, and may vary over time (Judd & Swanstrom, 1988; Shefter, 1985; Vogel, 1992). Second, these political arrangements do not afford land-based businesspeople with the kind of controlling power over local politics depicted in the earlier elitist analysis of the city (see Domhoff, 1978). The more recent urban political-economic analysis, applying Lindblom's general model of state-market relations in capitalist democracies, finds these interests to be, instead, *privileged* (cf. Lindblom, 1977, 1982; see also Block, 1987, chap. 3; Elkin, 1987, p. 46). The political voice of this group, though not the only one audible, is unquestionably the loudest. More concretely, this alliance provides the land-based business community a degree of access and influence unmatched by other political interests in the city (cf. Stone, 1989).

Seen in this light, the problem confronting "complex" political equality is readily discernible: The members of the land-based business community are able to *convert* their institutional position in the economic sphere of the city into substantial power in its political institutions. Using Walzer's terminology, one can say that in contemporary U.S. central cities, separations don't hold—thus violating the autonomy of the political sphere. The result is political inequality among the urban citizenry.

The Problem of Economic Inequality

Some measure of economic inequality in a democracy is no doubt generally consistent with the demands of complex political equality. "There are, of course, constraints and inequalities within each institutional setting," Walzer (1984) explains, "but we will have little reason to worry about these if they reflect the internal logic of institutions and practices" (p. 321). And the "internal logic" of economic institutions and practices in a liberal-democratic society clearly generates material inequalities among citizens, as the strong protection of individual liberties allows for substantial private control of productive assets and places significant limits on the socialization of consumption (cf. Elkin, 1987, pp. 184-188; Gutmann, 1980). As Dahl demonstrated long ago, however, citizens have access to a variety of political resources in a liberal-democratic society; thus, a skewed distribution of income—in and of itself—need not produce patterned bias (see Dahl, 1961, especially chaps. 19-23).

Nevertheless, within most contemporary American central cities, economic inequality *is* a formidable obstacle to the achievement of complex political equality. This conclusion is informed by the considerable body of research confirming the emergence of an urban "underclass"—a permanent substratum of citizens locked out of mainstream American life and suffering severe economic deprivation (see Wilson, 1987, 1989).[10] Although the actual size of the urban underclass is not immense—upper-bound estimates are in the 2.5 million range (Fainstein, 1993, p. 344)—the existence of this segment of the urban population substantially threatens the democratic character of the central city (cf. Kaus, 1992). In short, given the intensity of the economic deprivation experienced by these citizens, the specified criteria for political equality cannot be met in many urban communities.

Whereas we traced the first impediment to complex political equality —the problem of privileged voices—to the ability of some citizens to convert *success* in a nonpolitical sphere into considerable political power, the impediment here arises from the deep-rooted *failure* of some citizens in a nonpolitical sphere and the political disempowerment it creates. In essence, certain inequalities in the urban setting are *cumulative:* The underclass population of the city is so disadvantaged economically that this translates into extreme political disadvantage. A deep and incessant poverty brought on by chronic unemployment and acute educational deficiencies—along with behavioral patterns deviating from prevailing social norms and attitudinal problems of isolation and alienation[11]—undercuts even minimal requirements for equal democratic citizenship (cf. Alex-Assensoh, 1995; Cohen & Dawson, 1993; Howard, Lipsky, & Marshall, 1994, pp. 189-190; Shaw, 1994).

That the policy agenda pursued by *local* governing coalitions has exacerbated substantially the plight of the urban underclass is an often overlooked point. Nevertheless, consider the following causal connections.

First, as Wilson's (1987) influential and comprehensive research on the topic emphasizes, a crucial factor in the development of the underclass has been the "social isolation" of the urban minority poor—the spatial separation of this population from the people and institutions of mainstream society (p. 60; see also Goldsmith & Blakely, 1992, p. 136).

Concentrated in impoverished ghetto neighborhoods, these citizens lack frequent contact with stable working persons. This circumstance in turn breeds patterns and norms of behavior among ghetto dwellers inconsistent with the development of work habits necessary to obtain steady employment. The overall result of this protracted unemployment and underemployment is, according to Wilson and others, a condition of extreme economic deprivation among inner-city minorities.

For Wilson, the key to this social isolation is the loss of the "social buffer" in these neighborhoods, which occurred when higher-income minorities moved away from the inner city after the passage of civil rights legislation. Yet as the political scientist Adolph Reed (1988b) points out, another major source of social isolation was the spatial effect of the corporate-center/mainstream pro-growth land use decisions made during the period of urban renewal. These policies, Reed (1988b) writes, "cut off minority communities, displaced large sections of these communities and concentrated them between expressways, office complexes, stadiums and civic centers" (p. 169; cf. Friedland, 1983; Harvey, 1973; Stone, 1976).

A second major factor contributing to the development of the underclass has been the structural change of the urban economy "from centers of production and distribution of material goods to centers of administration, information exchange, and higher-order service provision" (Kasarda, 1985, p. 33; see also Kasarda, 1989, pp. 28-33; Wilson, 1987, pp. 39-46, 100-104). This shift altered the nature of employment opportunities in central cities: "Knowledge-intensive" jobs, requiring a high degree of skill and education, increasingly have replaced jobs in manufacturing and other blue-collar industries that required only limited formal training. Because a large segment of the urban minority population lacks the level of schooling necessary to obtain knowledge-intensive employment, a "mismatch" has developed between the skills of central city residents and the educational and training requisites of existing urban employment opportunities. The results of this jobs-skills mismatch, like the condition of social isolation discussed above, are widespread and persistent inner-city minority joblessness and the concomitant emergence of the urban underclass (Kasarda, 1985; Wilson, 1987).

This tragic reality has been exacerbated, once again, by the policy priorities of local governing coalitions. Surveying this question, one scholar (Levine, 1988) points out that, rather than "reinvesting in revitalized public education so that city residents might be better equipped to participate in an information-based economy" (p. 125), urban officials allowed the jobs-skills mismatch to expand in an unimpeded fashion. Likewise, as Stone et al. (1991) note, local efforts at "human capital enhancement" have been relatively meager, while at the same time cities have devoted considerable energy and resources to the pursuit of corporate-center/mainstream strategies for urban economic growth, especially via the physical redevelopment of the central business district: "Policymakers have adjusted to the changing social economy of the city by modifying land use in an effort to spur investment within downtown areas," they explain. "The building of office towers, hotels, convention centers, exhibition halls, festival marketplaces, sports facilities and transportation networks has been the predominant concern of city leaders" (p. 222).

Pursuing this agenda has not only diverted local energies and resources away from the kind of human capital enhancement efforts that might ameliorate the mismatch problem. Beyond these "opportunity costs" (see Riposa & Andranovich, 1988, p. 33), evidence strongly suggests that these policies themselves have proved to be ineffective as a means of improving the lives of urban poor: As Elkin (1987) concludes after evaluating these efforts, "There is good reason to believe that [these currently employed] growth strategies themselves contribute to inequality" (p. 100; see also Barnekov & Rich, 1989; Clavel & Wiewel, 1991, pp. 5-6; Fainstein et al., 1983; Ganz, 1986, p. 53; Krumholz, 1991; Leitner & Garner, 1993, pp. 64-65; Stone et al., 1991, pp. 222-223; Weiher, 1989, p. 233). The deleterious effects of land use patterns associated with this approach to urban economic development already have been mentioned (see Reed, 1988b). In addition, as Marc Levine (1987, pp. 116-118) points out in his evaluation of the Baltimore experience, the corporate-center strategy pursued by that city and elsewhere lacks the means to channel downtown growth so that it stimulates economic activity in distressed neighborhoods. As a result, "in city after city, redevelopment has been associated with a 'tale of two cities': pockets of revitalization surrounded by growing hardship" (Levine, 1988, p. 124).

In summary, then, evidence suggests that the policy agenda pursued by central cities, though not exclusively responsible for the current plight of the urban underclass, has contributed to it in notable ways. As a corollary of this fact, it follows that a modification of the urban policy agenda in the appropriate directions, if achievable, would serve as an efficacious means of mitigating some of the worst elements of economic—and *hence political*—deprivation experienced by this group of citizens (cf. Goldsmith & Blakely, 1992, pp. 180-183; Robinson, 1989, pp. 283-284; Stone et al., 1991, pp. 232-236).

LOCAL POLITICAL INEQUALITY AND THE LARGER LIBERAL-DEMOCRATIC ORDER

This chapter began with the idea that the study of city politics should not be undertaken in isolation from the larger political order of which it is part (see Elkin, 1987, chap. 1). Specifically, the general inquiry revolves around the question of what the reconstruction of central city politics contributes to the promotion of one desirable political way of life—liberal democracy. Therefore, a pause must be taken in the development of the argument so as to establish the *connection* between the realization of increased political equality in central cities and the flourishing of the larger liberal-democratic order.

Many great theorists of liberal-democratic regimes from Alexis de Tocqueville to Robert Dahl to James Buchanan have recognized the need for the institutionalization of decentralized public authority to protect individual liberties and realize popular sovereignty. Lodging too much power at a centralized level, these theorists argue, breeds tyranny and undermines democracy, as government becomes too distant to be responsive to individual preferences, and the political unit grows too large to encourage effective and meaningful citizen participation (cf. Dahl, 1967; Hambleton, 1990, p. 89). Hence the decentralization of state power stands as an important institutional buttress for the maintenance of the liberal-democratic political way of life.

These theorists face a dilemma, however: As I have described earlier, the current state of political life in some of the local jurisdictions where power is decentralized—central cities—is plagued by systemic bias (cf. Elkin, 1987; Stone, 1980). As a result, the devolution of state power in this context fails to foster the values of the larger liberal-democratic order: Although localized, the existing systemic bias nevertheless still leaves public authority structured in ways making it both unresponsive to individual citizen preferences and inhospitable to meaningful citizen participation. Decentralized state power, it might be said, must be properly constituted.

The reform of urban politics outlined above to increase the (liberal) political equality in central cities is a response to this dilemma: Protecting the integrity of the local political sphere from encroachment counteracts systemic bias. Briefly stated, once the workings of urban politics are altered in the necessary directions, decentralization can provide the sustenance for a healthy liberal-democratic order.

Yet what about decentralization itself? It may be vital for the larger political order, but how is it to be established? While this critical task at first appears arduous, there is reason to believe that it is currently being realized. As explained below, a wealth of research produced by scholars associated with the regulationist school indicates that powerful forces at work on a global scale are currently devolving political power to local units.

These forces emanate from the ongoing process of global economic restructuring—in particular, what regulation theorists identify as the transition from a Fordist system of accumulation, based on mass production, thriving from the 1920s to the 1970s, to a post-Fordist system of accumulation based on flexible systems of production (cf. Aglietta, 1976; Lipietz, 1986; Piore & Sabel, 1984).

For regulationists, each system of accumulation, to be viable, requires a corresponding mode of social regulation—"a web of complementary social phenomenon [that] comes into being alongside . . . [a system of] accumulation as a means of stabilizing its operation through time" (Scott, 1988, p. 172). Fordism achieved regulation through the institutions and norms of the centralized Keynesian welfare state, which buttressed the mass consumption required to avoid a crisis of overproduction brought on by Fordism's massive productivity gains. Alternatively, with the crisis of Fordist accumulation in the 1970s in both North American and Western Europe "there has been a wholesale dismantling of the apparatus of Keynesian welfare-statism" (Scott, 1988, pp. 174-175), and the subsequent development of a new mode of regulation, marked by consumption patterns more differentiated and specialized, which corresponds to and supports the system of flexible accumulation.

The essence of this new mode of regulation is the decentralization of state power (Logan & Swanstrom, 1990, p. 12; Mayer, 1988; see also

Sabel, 1982, pp. 228-229). Sketching the implications of this transformation for the "post-Fordist city," Margit Mayer (1988) notes that the
breakdown of Fordist accumulation and its centralized mode of social
regulation embodied in the Keynesian welfare state "has conferred new
challenges on the regional and local levels and ideologically upgraded
local politics everywhere . . . opening the door to a type of municipal
autonomy not known before" (p. 3; 1991, p. 112; cf. Clavel & Kleniewski,
1990, pp. 225-227). In particular, Mayer (1988) describes the local level's
emergence as "an essential instrument in the establishment of flexible
accumulation," (p. 23) and cites several examples of the "current devolution of accumulation-related state functions" (p. 27) to cities (see also
Mayer, 1991, p. 116).[12] In short, the actions of city and regional governments increasingly determine a nation's economic performance and
distributional outcomes (see Clarke & Gaile, 1992, p. 188; Fosler, 1988,
pp. 5-8; Haider, 1992, pp. 133-134; Osborne, 1988; Preteceille, 1990).[13]

Many observers of the contemporary urban condition remain less
sanguine about the prospects for the city in the post-Fordist era, however.[14] These analysts view the intensity of interjurisdictional competition for capital associated with the new flexible mode of regulation as
extremely pernicious to the urban polity (see Gottdiener, 1987, especially chap. 1). Similarly, as I have argued, this new decentralized power
is, in the terminology used above, now improperly constituted (or
systemically biased)—and therefore must be reconstituted. Yet this
conclusion does not detract from the essential point: In the current era
of post-Fordism, subnational (i.e., decentralized) governments increasingly undertake important matters (defined as those significantly affecting the well-being of citizens, positively or negatively). Hence, the
necessary devolution of political authority vital to the larger liberal-
democratic order is clearly discernible.

THE CENTRAL CONSTRUCTIVE TASK

The analysis in this chapter traces the extreme political inequality that exists in the central city to the *constitution of contemporary urban regimes*—that
is, to the makeup of local governing coalitions and the nature of the policy
agenda pursued by these actors. The central constructive task is now
evident: The design of institutional settings in central cities must be
altered to promote the reconstitution of current urban regime forms.

Accomplishing this central task requires, first, an inquiry into the reasons why these urban regimes form as they do. This matter will be explored in the chapter to follow. Before proceeding in that direction, however, I will briefly describe the essential features of these reconstituted urban regimes.

First, consider the current state of political configurations in central cities. As previously argued, political inequality stems from the close alliance between public officials and land-based business interests at the heart of most local governing coalitions. The key to reconstituting this first element of urban regimes is to somehow break or significantly weaken this alliance and, as a result, to promote the formation of alternative governing coalitions. Put directly, in the politics of the city the political concerns of businesspeople must continue to carry weight (cf. Elkin, 1987), but their level of influence needs to be placed on a more equal footing with political actors representing community, nonprofit, and neighborhood groups. These reforms amount to, in short, what Stone (1989, p. 243) labels the process of making urban regimes *more inclusive*.

As also previously argued, the second feature of the dominant urban regime form—the orientation in local political agendas—contributes to the existence of political inequality in central cities through its role in perpetuating the extreme economic deprivation of a portion of the urban citizenry. At a general level, the reconstitution of this facet of urban regimes involves expanding the city's agenda to include not only growth policies but also social policies broadly redistributive in nature. In particular, much more energy and additional resources need to be devoted to policies designed to enhance "human capital" (e.g., educational reform, job training, placement efforts). In short, the general model, as Stone et al. (1991, pp. 232-236) specify, is to create an urban regime having as its main policy task the goal of *expanding opportunities for the lower class in cities*. This agenda, in turn, mitigates the worst aspects of the extreme economic inequality in the city, furthering the degree of political equality in the city.

Reconstituting these two central elements of urban regimes in this manner will not solve all of the problems currently afflicting American central cities. Nevertheless, creating a less biased local political system by increasing political equality improves the quality of urban democracy and, as a result, places the larger liberal-democratic order on a more solid foundation.

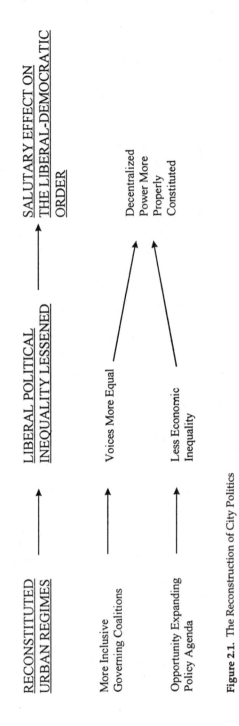

Figure 2.1. The Reconstruction of City Politics

LOOKING AHEAD

Now the argument presses on to consider the questions of *why* and *how:* why current urban regimes form as they do and how these regimes might be reconstituted.

Regarding the why question, I demonstrate that two broad structural features of the urban polity leave city public officials in a dependent position. These dependencies, in turn, are the key causal factors explaining current urban regime formation. In addressing the question of how to reconstitute such regimes, I explore the potential embodied by a reorientation of city economic development strategies in an "alternative" direction. This exploration—conducted via the presentation of an array of empirical evidence and case analysis—preoccupies the balance of this book.

NOTES

1. For a representative sample, see Cummings (1988), Elkin (1987), Fainstein et. al. (1983), Friedland (1983), Gottdiener (1987), Jones and Bachelor (1986), Kantor (1988), Logan and Molotch (1987), Molotch (1976), Peterson (1981), Shefter (1985), Stone (1989), Stone and Sanders (1987), and Swanstrom (1985).

2. Although rare, other types of urban regimes do exist. For example, Clavel (1986) describes instances of "progressive" patterns in local politics. Yet Rosdil's (1991) study of urban regimes exhibiting this pattern suggests that, in the present urban political-economic context, they are likely to remain confined to those few cities with a particular cultural and economic makeup (see also Magnusson, 1989).

3. Traditionally, key actors included banks, newspapers, large stores, developers, real estate agencies, and utilities (Elkin, 1987, p. 41). Large multilocational corporations with long-standing ties to, and large sunk investments in, cities also can be added to this list (cf. Stone, 1989, p. 169). More recent trends in economic restructuring have, however, reduced the importance of some of these traditional players, for example, large downtown department stores. In addition, the recent wave of corporate restructuring (mergers, buyouts, etc.) and the increasingly mobile nature of capital in the postindustrial era make multilocational corporations less important than in the past (see Jones & Bachelor, 1986; Kantor, 1988).

4. In her survey of more than 300 cities, Bowman (1987b, p. 8) reports that mayors listed economic development as one of their top priorities more often than any other policy activity (e.g., public safety, education, housing, transportation).

5. Stone (1989) defines a city's governing coalition as "a core group—typically a body of insiders—who come together repeatedly in making important decisions" (p. 5). The governing decisions of this coalition, it needs to be emphasized, are, as he points out, "not a matter of running or controlling everything. They have to do with *managing conflict* and *making adaptive responses* to social change" (p. 6).

6. Its membership is not necessarily confined to these actors, however. Less central members may include labor-union officials, party functionaries, officers in nonprofit organizations or foundations, church leaders, and other community-based actors (see Orr, 1992, p. 185; Stone, 1989, p. 7).

7. For some examples of this account, see Mansbridge (1977, 1980) and Gutmann (1980). "By political equality," Jane Mansbridge (1977) writes, "I mean equality of power" (p. 321). For reasons different from the ones offered below, Beitz (1989, pp. 4-8) also rejects this "simple view" of political equality.

8. Of this disparity in political power, Walzer (p. 304) notes, "Individual citizens always share in decision making to a greater or a lesser degree. Some of them are more effective, have more influence, than others. Indeed, if this were not true, if all citizens had literally the same amount of influence, it is hard to see how any clear-cut decisions could ever be reached" (p. 304).

9. That is, by this standing as opposed to the persuasiveness of their arguments. Walzer (1983) argues instead the "rule of reasons" must prevail: "Citizens [must] come into the forum with nothing but their arguments. All nonpolitical goods have to be deposited outside: weapons and wallets, titles and degrees" (p. 304). On a different point, Walzer (1983, 1984) also considers as a violation of "complex equality in the sphere of politics" the opposite possibility—that those with access to political power can gain superior standing in economic, social, or religious institutions.

10. The term *underclass* has been attacked as contributing to the stigmatization of the urban poor (see Gans, 1990; Hochschild, 1991). In response, Hochschild, for example, substitutes the term *estranged poor*. Though Wilson used underclass in his earlier work (1987, 1989), he now prefers the term *ghetto poor* (see Swanstrom, 1993, p. 68). Though I am aware of the undesirable baggage associated with the term underclass, this work will still use it, because any alternative term—once widely employed—might soon also be seen as stigmatizing.

11. I derive these comments concerning the plight of the underclass from the excellent summary provided by Nathan (1989, pp. 171-172).

12. Such as cities "imposing [development] controls on . . . [their] . . . physical environment" and "making stipulations regarding labor conditions and employment guarantees. . . . In the framework of the Fordist mode of regulation," Mayer (1988, pp. 26-27) adds, these matters "used to be centrally and uniformly codified." More generally, Mayer (1991) continues, under the new globalized economic conditions, "since it is impossible for the central government to organize the variety of conditions that each [production] location requires, local governments have become responsible for negotiating directly with super-regional (and multinational) capital and for tailoring environments to meet companies' needs. . . . The result has been a decentralization of investment-related decision-making power." (p. 116)

13. As Michael Piore and Charles Sabel (1984) conclude in their influential study of flexible accumulation, "In our analysis, successful industrial reorganization in the United States will require the reinvigoration of local and regional government" (p. 301). Similarly, Clarke and Gaile (1992) speak of the recent emergence of a "new economic localism" (p. 188).

14. Mayer herself is ambivalent (Mayer 1988, pp. 3-4, 27; 1991, p. 123). See also Clarke and Gaile (1992, pp. 187-188).

CHAPTER 2

RECONSTITUTING URBAN REGIMES

EXPLAINING URBAN REGIME FORMATION

Reconstituting current urban regimes requires understanding *why* these political arrangements develop as they do. This inquiry allows for an isolation of the critical practices and institutional relationships that shape the urban regime formation process. Once these practices and relationships are isolated, we can then understand how they can be modified to promote the desired reconstitution.

To reiterate, general agreement exists in the urban political economy field about the fundamental features of contemporary central city politics in the United States. Two closely related empirical tenets command wide support: (a) that there is often a close governing alliance between public officials and the local land-based business community and (b) that this alliance is preoccupied with a policy agenda oriented toward promoting local economic growth via corporate-center/mainstream economic development strategies. Nevertheless, this consensus on the nature of urban regimes—or what I have termed the *dominant urban form*—exists at a descriptive level only. In the realm of explanation, urban political economists have advanced widely contrasting accounts of why urban politics looks as it does, that is, why urban regimes often form in the manner outlined above.

Simplifying matters slightly yields two broad perspectives on the subject. These alternative models of urban regime formation enliven the so-called "city limits" debate.[1] Although typically cast as an argument about the explanatory primacy of social structure versus human agency, and/or the relative influence of economic variables versus political variables, neither characterization of this well-known debate is entirely appropriate. The first characterization misleads us because, as shown below, both perspectives in the debate rely heavily on structural factors to explain urban political phenomena. The second characterization, while helpful, needs to be further specified. To clarify exactly what is at issue, the dispute should be understood as a contention between explanations grounded in the workings of either the *exogenous (economic) environment faced by cities or the endogenous (political) dynamics within cities themselves.*[2]

The External Economic Explanation

Paul Peterson's (1981) familiar argument in *City Limits* best exemplifies the first view—the perspective focusing on external economic forces as the determinants of urban regime formation, but it also is represented by the writings of numerous other urban political economists working from a broad array of theoretical paradigms.[3] From this standpoint, two features of the external environment are particularly consequential for American central cities. First, the global economic restructuring in the postindustrial era—especially the rise of the multilocational corporation as the dominant organizing force in local economies—has left economic wealth less wedded to particular urban locations (Friedland, 1983; Jones & Bachelor, 1986; Kantor, 1988, pp. 164-170). Most important here is the increasing mobility of investment capital, and the corresponding legal prohibitions preventing local public authority from controlling these capital flows (Peterson, 1981, pp. 28-29). A second external feature, in turn, complements this circumstance: the existence of a relatively decentralized federal system that encourages economic competition among subnational governmental units, as these individual jurisdictions possess considerable autonomy to pursue contrasting policy objectives.

These two external factors—the increasing mobility of capital and the decentralized federal system—are seen as making cities extremely vulnerable to economic disruptions brought on by capital disinvestment. And because local governments receive little financial assistance from other governments, they must rely heavily on revenues collected from sources rooted in the local economy to pay for city services and to keep the overall local tax burden relatively low. Thus, the fiscal integrity of urban governments, along with "employment opportunities, income, and the general well-being of city residents" is, as Shefter (1985) notes, "tied to the vitality of the local economy," (p. 5) and that vitality is contingent, most decisively, on the ability of cities to attract and retain mobile investment within their borders. Consequently, the failure of city public officials to accomplish this crucial task threatens their political futures.

How, then, according to this external perspective, is the pervasiveness of the dominant regime form in urban America to be explained? Consider both of the constituent elements of urban regimes—the governing alliance between local public officials and land-based business interests and the dominance of corporate-center/mainstream development policies on the urban agenda.

First, this perspective understands the latter as a rational policy response by urban public officials to the strong economic pressures cities confront (see Peterson, 1981). These officials must, by necessity, heed external imperatives or face financial ruin, widespread local unemployment, and a decline in the standard of living of the city's inhabitants— all outcomes that are, as noted above, likely to undermine their legitimacy. The tautness of the economic constraints leave little political choice in the matter: Luring holders of capital assets into the city and preventing others from leaving demands that urban public officials act in certain ways; namely, they must pursue an agenda focusing heavily on corporate-center/mainstream economic development policies.

This perspective traces the second element of contemporary urban regimes—the development of close governing alliances between public officials and segments of the local business community—to the same source. In essence, the economic constraints on cities compel public officials to ally themselves closely with business interests so they can respond successfully to the powerful growth imperatives plaguing the city. Because cities are locked in a competitive economic struggle with other localities, their political leaders cannot afford to be timid about cultivating this relationship (cf. Cox & Mair, 1988). Doing otherwise would jeopardize the city's chances to pursue economic development and thus could cost local elected officials substantial public support (see Peterson, 1981, pp. 146-149).

The Internal Political Explanation

Other urban political economists, in contrast, contend this account of urban regime formation, with its focus on external economic structure, is inherently flawed. They argue, instead, that regime patterns only can be accounted for through an appeal to the internal political dynamics of the city. Summarizing the essence of this perspective, Swanstrom (1988) writes, "Land interests, in coalition with growth-oriented politicians, dominate the agenda of city politics. But this is more for [internal] political than [external] economic reasons" (p. 107).

Although this second model acknowledges that external economic constraints faced by the city are real (Stone, 1987, pp. 12-16), these factors serve basically as background variables, setting the general political-economic context for urban political explanation but not in themselves revealing much about the specifics of regime formation.[4] This conceptualization is advanced because it is believed that the translation of external economic pressures into the local policy-making arena is not a frictionless process. Internal political competitions and structures mediate strongly between those pressures and urban political outcomes (Swanstrom, 1988, p. 107). As a result, this view sees the composition of the city's governing coalition and the orientation of the city's agenda—that is, the nature of its urban regime—primarily as a reflection of the forces at work within the internal urban political system. To quote the mantra adopted by the advocates of this position, in explaining urban regime formation, "Local politics matters" (cf. Waste, 1993).

Consider first the ubiquitous allure of corporate-center/mainstream growth policies to urban public officials. From this second perspective, this allure does not stem from the rational reactions of these officials to the external constraints on the city. Instead, the aggressive pursuit of this agenda is thought to be attractive for its political rather than economic advantages. In John Mollenkopf's (1989) words, such actions are "a natural way of cementing a coalition among . . . otherwise disparate elements" (p. 132) in the urban political community (see also Mollenkopf, 1983). Of crucial importance for this practice—which has come to be known as "growth politics" (Elkin, 1987; Logan & Molotch, 1987; Swanstrom, 1985)—is that these growth policies usually generate a stream of selective benefits that can be manipulated by politicians to build internal political support (Stone, 1987, p. 11). There is, for example, a powerful nexus between these efforts and the flows of development opportunities, money, contracts, and jobs that can be targeted to specific individuals and companies and easily translated into campaign contributions and other election benefits (Elkin, 1987, p. 38; Logan & Molotch, 1987). Here lies the real seduction of this growth agenda in contemporary urban politics, adherents of this second view argue.

In fact, as understood from this perspective, the pursuit of urban growth itself by local public officials rarely, if ever, yields the fiscal and economic dividends promised by its advocates (see Barnekov & Rich, 1989; Feagin, 1988; Friedland, 1983; Molotch, 1976; Stone, 1987). Although "the costs and benefits of growth depend on local circumstance," Logan and Molotch write (1987), "for many places and times, growth is at best a mixed blessing. . . . Residents of declining cities, as well as people living in more dynamic areas, are often deceived by the extravagant claims that growth solves problems" (p. 85). This view argues that although urban growth may bring an increase in tax revenues for the city, this growth also increases demand for urban services and often requires extensive investment in local infrastructure. Furthermore, many of the potential benefits from growth—including most notably the creation of quality employment opportunities—often "leak" to the surrounding suburbs. The general claim advanced, then, is that the net effects of this increased economic activity on the urban community are usually negative. As a result, from the internal political perspective, explaining the current constitution of urban policy agendas via an appeal to the rational reactions of local officials to external economic pressures faced by cities is fundamentally misguided.

Likewise, the internal view also minimizes the role played by external imperatives to explain the other characteristic of the dominant urban regime form—the key position of the land-based business com-

munity in governing coalitions. This perspective attributes the prepon-
derance of influence wielded by business to the substantial body of
political resources these interests hold and use within the internal urban
political community, relative to the resource capacity and usage of other
interests. The resulting patterns of influence in the city stem from the
internal struggle for political power—not the exigencies of interjurisdictional economic competition (see Stone, 1989).

That competition, proponents of the internal view further argue, is
not as acute as is often claimed. There are powerful limits on capital
mobility, for example. The large sunk investments in particular cities
held by many companies and the existence of place-specific business
opportunities in urban settings cause much capital to be anchored in
particular locations (Stone, 1988). Moreover, although public officials
need capital, controllers of capital need public officials. To prosper,
businesses often require the local state to provide the necessary conditions for profitability. Economic dependence is not unidirectional; bar-
gaining occurs, and adjustment is mutual.

In short, the following counterfactual statement captures the essence
of this internal perspective: Even if external economic pressures were
greatly reduced, the attraction between public officials and land-based
business interests for the most part would remain, and city politics and
public policy would look much as it does now.

Beyond the "City Limits" Debate: Rethinking Regime Formation

Reflecting on the current state of this city limits controversy, Jeffrey
Henig (1992) points out in a recent paper that "as with the community
power debate, there is a danger that initially fruitful theoretical ex-
changes may peter off into [a] stale stand-off" (p. 376). Similarly, another prominent urban scholar, Bryan Jones, writing in the American
Political Science Association's *Urban Politics Section Newsletter*, assesses
the recent developments in the urban political economy field and makes
reference to the potential "cul-de-sac" faced by the debate's antagonists.
In light of this intellectual impasse, I want to suggest that any adequate
model of urban regime formation must be built from the premise that
both external economic factors and internal political factors play a
critically important role in the process.

As a first step toward building this integrated theoretical model, consider the following alternative conceptualization of the problem.

As stated earlier, the key tenet of the endogenous political perspective is that external economic pressures do not play a powerful role in urban regime formation because the local political context forcefully mediates these pressures. This formulation implies, correctly, that the distinct internal political relationships, competitions, and structures found in different cities act to alter significantly the nature of the external economic imperative. The specific face of local politics—which intervenes (or mediates) between economic pressures and the formation of urban regimes—is, thus, the crucial causal variable; consequently, it explains much about urban outcomes. Nevertheless, we must understand that although political arrangements alter the nature of economic pressures (see Horan, 1991, pp. 119-120), the converse is also the case: Economic pressures modify the nature of the political arrangements themselves. The prevailing pattern of local political arrangements is, then, *both subject and object*. It mediates economic pressures, but the nature and extent of these pressures strongly affect its own development. Hence, this vantage point incorporates the essence of both perspectives into a single theoretical model: How city politics looks is the result of a city's internal political struggle (a point omitted from the external economic perspective), but that struggle is itself deeply influenced by the extent and nature of urban economic imperatives (a point often underemphasized in the internal political perspective).

What about the specific elements of the dominant regime form? How can their existence be reinterpreted in this new light? I want to suggest that the attraction of urban public officials to the local land-based business community in the city's governing coalition (and, ultimately, the pervasiveness of the corporate-center/mainstream growth agenda) can be fruitfully conceptualized as resulting from the "dual dependency" of public officials. Given the current structural conditions of the urban political economy, those wielding public authority require both (a) the attraction and retention of external capital investment and (b) access to resources situated in the internal political community. As a result, urban public officials face *both an external economic dependence and a internal resource dependence*. These dependencies, I shall argue, are, at root, the causal determinants of the dominant regime form in contemporary central city politics.

External Economic Dependency

Earlier we understood external economic dependence as the need for local public officials to attract and retain sufficient amounts of external economic investment (i.e., either prospective investment capital or capital currently located in a city but relatively mobile). To reiterate, the essence of the relationship at work here involves the nexus between good local economic performance and the continued ability of local officials to get elected. Accomplishing this central task—achieving the necessary levels of urban development—is thought to be demanding for public officials in most jurisdictions, given that they usually face an environment marked by intense interlocal economic competition (see Blair & Wechsler, 1984; Bowman, 1988).

As noted above, many urban scholars downplay the role of external economic influences as a driving force behind the development of urban regime patterns. Yet these scholars assume this theoretical posture because of flaws found in recent accounts of city politics that use external economic influences as an important explanatory variable. Two of these flaws stand out.

First, the analysis provided in these studies is often unidimensional and deterministic: The economic dependence of the city carries all of the burden of explanation in the analysis, leaving no room for autonomous political conditions to play a nontrivial role in regime formation. In contrast, as will be seen below, the model of regime formation developed here acknowledges that internal political factors also play a major role in the process.

Second, these studies often erroneously conceptualize the effects of economic pressures on cities. Correcting for this flaw, the model presented here recognizes that economic constraints foster the development of *political arrangements;* they do not—in and of themselves— necessitate a particular kind of urban policy response.[5] Contrary to the ideas of Peterson (1981) and like-minded urban theorists, this model posits that within broad parameters cities have choices about how they will react to their external environments (Kantor & Savitch, 1993). These choices, however, are made via the political arrangements constituting local regimes; arrangements biased—due *in part* to the effects of economic pressures on cities—toward the interests of the land-based business community. Todd Swanstrom's (1986) insight seems apt here: Even though cities have economic space to maneuver (insofar as economic pressures do not imply a *particular* policy response), the political space is much more constrained. Yet as just noted, the two are intimately connected—notably, the economic pressure on the city contributes to the compression of this political space.

Moreover, the attraction of urban public officials to business interests is *not* being portrayed here as simply a case of the public sector courting capital to prevent its mobility. The dynamic at work is more complicated: As described above, those business interests constituting local regime alliances *are tied* to a considerable degree to the local economy. Their capital investment, actual or prospective, is relatively immobile (see Elkin, 1987, pp. 40-41).[6] Instead, the forces emanating from external economic dependence function as a catalyst impelling local public officials to build close political alliances with these interests. In essence, local public officials require the attraction and retention of *mobile* capital, and this leads them to seek an alliance with local land-based (*immobile*) capital.

In the face of external economic dependence, public officials perceive this alliance to be valuable for a number of reasons. First, land-based businesspeople in the local community and the organizations they control possess technical expertise and information concerning development matters thought by officials to be indispensable for promoting the local economy (see Clarke, 1987). Without this expertise and knowledge, city officials feel handicapped in their competitive economic struggles with other communities. Second, to attract and retain capital investment in their cities, public officials often attempt to enhance the image of their communities—their so-called business climate—as it is projected to prospective investors. Maintaining a close and accommodating association with the local business community is seen as an effective way of acquiring the proper reputation.[7] And finally, public officials also find in the land-based business community a creature equally eager to see external investment capital find its way into the city—a "natural ally" of sorts. Owning large fixed assets tied to specific locations, these interests seek to enhance the value of their investments via the intensification of city development wrought by the attraction or retention of mobile capital (see Cox & Mair, 1988; Humphrey et al., 1989; Logan & Molotch, 1987). Among the assorted interests in the city's political life, only these actors possess a concern for economic growth matching (or even exceeding) that of the local public officials. Hence, land-based business interests can be expected to devote considerable energy toward assisting public officials in their efforts to respond successfully to the economic dependence of the city.

Internal Resource Dependency

The dependence of public officials on attracting external capital in part explains the development of urban governing arrangements that revolve around an alliance between local public officials and land-based business interests. Yet a second broad structural process, one firmly rooted in the internal *political* dynamics of the city, also contributes to the emergence of these patterns. In the external economic dependence of the city, the key structural dynamic was heightened economic competition fueled by the increasingly mobile nature of investment capital. In contrast, the key dynamic at work here is the inherent weakness of the formal authority and institutional capacity of the local state in the face of an increasingly complex urban environment.[8] Most significantly, the local state's regulatory and fiscal powers are limited, and its apparatus has been splintered by the proliferation of special districts and public authorities.

This context limits the ability of public officials acting autonomously to accomplish the complex policy tasks required for effective urban governance. Instead, carrying out these tasks demands that these officials meld their limited resources with the resources held by organizations in the larger urban community so as to create a capacity to act in the local polity (Stone, 1986, 1989). Urban policy effectiveness—not only in economic development but in such realms as public safety, education, housing, transportation, health, and arts promotion—often requires that local public officials obtain the cooperation and support of the resource-laden entities in local civil society. They experience, in short, what might be called an "internal resource dependency."

Creating the effective capacity to govern is of momentous importance for local public officials, especially those politicians faced with an election cycle. As Elkin (1987) notes, city officials, not unlike officials at other levels of government, "believe that their electoral prospects are markedly improved if they can secure a reputation for promoting innovative policies and if, in general, they are associated with publicly visible activities of almost any sort" (p. 37; cf. Stone, 1980). However, when officials lack the ability to act effectively in their communities (i.e., when local governance is weak), this denies them the political advantages stemming from their association with these innovative and visible activities. This situation, in turn, can directly threaten these officials with electoral pressures or, at a minimum, severely limit their opportunities to win higher elected office.

Yet why are these "resource dependent" public officials attracted to the local land-based business community? A look at the distribution of resources within most urban communities clearly reveals the secret of this attraction. In the diffuseness and incoherence of local polities, the business community usually is the only political force in the city holding a *concentration of resources* (Stone, 1989, p. 229). And as Stone notes, "In a world of diffuse authority a concentration of resources is attractive" (p. 229). In particular, it affords those who wish to bring about social change the means to do so.[9] By comparison, the resource base of neighborhood and other nonprofit groups is often meager and fragmented (Stone, 1989; cf. Henig, 1982; Jezierski, 1990). Given this setting, Stone (1989) continues, the land-based business community with its concentration of resources is "uniquely able to enhance the capacity of a local regime to govern" (p. 233). Moreover, its economic stake in the future of the city provides these actors with a powerful motivation to use its resources toward that goal (Cox & Mair, 1988; Molotch, 1976). As a consequence, the alliance with these land-based business interests (as opposed to others in the urban community) can be empowering for local public officials (Stone, 1989, p. 232).[10]

In sum, then, the nature of the distribution of internal political resources in cities can be specified as the second structural element accounting for the attraction of city officials to land-based business interests in urban regimes. In fact, evidence suggests that cities with internal resource distributions structured in other ways tend to have alternative regimes. For example, researchers find many progressive regimes (see Clavel, 1986) in those communities that have a large resource-rich but noncorporate middle class. In these cities, a nonbusiness resource base can be marshaled for effective governance, leaving public officials freer to form other alliances (cf. Stone, 1989, p. 228).

34

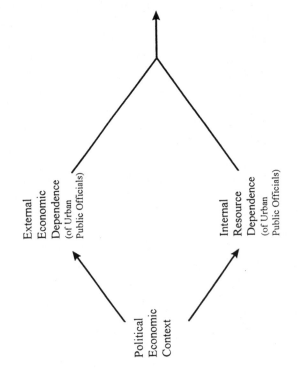

Figure 2.1. Model of Urban Regime Formation

"THE POINT IS TO CHANGE IT":
LESSENING THE DEPENDENCIES

Weakening the governing alliance between city public officials and the local land-based business community thus requires redesigning the institutional setting of the city in ways that alleviate the officials' "dual dependency."

With the weakening of this alliance, urban regimes can be reconstituted in ways that make them better approximate the normative ideal of (liberal) political equality. Both of the essential features—the who and what—of the more normatively desirable urban regime form (outlined at the close of Chapter 1) could emerge: First, these reconstituted regimes are likely to have *more inclusive* governing arrangements, involving a wider array of political interests in the city, hence redressing political inequalities. This result flows directly from the weakening of the alliance. Second, because of the intimate tie between the nature of the urban agenda and the character of a city's governing coalition (see Orr, 1992; Stone, 1987), once the alliance is weakened—and the governing coalition is made more inclusive—*policies more consistent with an agenda that expands opportunities for the lower class are likely to emerge.* This agenda, as noted earlier, tends to mitigate the political inequalities in the city that stem from the existence of extreme economic inequalities in the city.

If the key to necessary regime reconstitution lies, then, in the reduction of the structural dual dependencies of city public officials, *how* in broad terms can this be accomplished? Lessening the first dependency is straightforward: External economic pressures on the city must somehow be diminished. Lessening the second dependency can, generally speaking, be pursued via two alternative paths: (a) The resources of local units of public authority can be augmented, increasing the autonomous capacity of public officials to govern their cities effectively; or (b) the distribution of resources in the city can be altered so that business interests no longer possess the distinct advantage they now do.[11]

In more specific terms, the intensity of these dependencies could be diminished in a number of ways (see Imbroscio, 1993). Yet in this book I explore the potential embodied by one particular vehicle: the pursuit by cities of various "alternative" urban economic development strategies (i.e., alternatives to the current mainstream or corporate-center approach). Below, I offer a sketch of three such strategies—what I label a community-based strategy, an entrepreneurial-mercantilist strategy, and a municipal-enterprise strategy—and explain how, if implemented on a significant scale, they can conceivably bring about the necessary diminution of the dual dependencies.

Alternative Urban Economic Development Strategies

Community-Based Economic Development. Community-based economic development includes a broad array of institutions owned and controlled collectively by the local "community." These institutions are usually nonprofit (third-sector) organizations with democratic procedures, often based at the neighborhood level (Bruyn & Meehan, 1987; Gunn & Gunn, 1991).

The general model includes community-oriented economic institutions to encourage the "social emancipation of land, labor, and capital from the competitive market" (Bruyn, 1987, pp. 8-12). Community land trusts acquire real estate in the community interest; worker cooperatives give employees ownership and managerial control of businesses; and community finance institutions (such as community credit unions) provide investment capital for local development. In addition, consumer cooperatives extend the model from the production side of the economic development process to the consumption side of that process (Bruyn, 1987, p. 15). The core institution, however, is the community development corporation (CDC). CDCs serve as the "crucial coordination agent" in the community-based economic development process (Bruyn, 1987, p. 16).

Entrepreneurial Mercantilism. A second alternative local economic development strategy has no definite designation, but can be best labeled "entrepreneurial mercantilism." It is "entrepreneurial" in that it entails the selected intervention into the market economy by the local state to promote the rapid formation and expansion of new (especially small) businesses in the city. Cities advance the creation of new businesses by, for example, establishing publicly funded venture capital pools to meet financing needs, buttressing management skills and supplying various forms of technical assistance by creating business incubators, and helping local businesses find and exploit new markets for their goods and services, which includes demonstrating how they can take advantage of opportunities to market products to the public sector (Schweke, 1985, pp. 4-5).

It is "mercantilist" because an important overall goal is to reorient the economic development process so as to maximize the benefits accruing to the *local* economy from that process. Along these lines, policies would be designed around the idea of reducing a city's need for imports (see Long, 1987; Morris, 1982a; Persky, Ranney, & Wiewel, 1993), and would work to encourage both (a) the local ownership of businesses, which would increase the likelihood that economic assets will not flee the city (see Alperovitz & Faux, 1984, p. 146; Fisher, 1988, p. 175) and (b) a high degree of local interindustry dependence, which would enhance the economic multiplier effects of local purchasing and procurement activities (cf. Jacobs, 1969; Meehan, 1987). Likewise, cities would pursue efforts to conserve local economic resources by increasing the efficiency of resource usage (see Morris, 1982b), as well as efforts to tap innovative, locally generated finance sources such as municipal pension funds (see Ferlauto, Stumberg, & Sampson, 1992).

Municipal Enterprise. A third alternative strategy entails *decentralized* public ownership (and public profit-making activity more generally)— that is, what might be called municipal enterprise (Cummings, Koebel, & Whitt, 1989; Frug, 1980, 1984; Lynd, 1989, p. 19). Cities currently own convention centers, public utilities, and mass transit systems (Garber, 1990, pp. 11-12). The objective of this strategy is to expand city ownership and profit making to nontraditional areas of economic activity.

Frug (1980, pp. 1150-1151; 1984, pp. 687-691) suggests, for example, that cities own and often operate banks, insurance companies, cable television systems, grocery stores, and other profit-making businesses (see also Garber, 1990, p. 11; Osborne & Gaebler, 1992, pp. 214-215). Alternatively, cities might accumulate public property such as land or capital facilities for the purpose of leasing it to private for-profit entrepreneurs—but with a provision for the retention of public ownership (see Lynd, 1989, p. 19; Clavel, 1986, pp. 30-36). Moreover, cities pursuing this strategy could use their investments of public resources in private business ventures to gain a public equity holding or a share of the profits in such ventures (see Cummings et al., 1989, p. 202; Lassar, 1990; Osborne & Gaebler, 1992, pp. 200-202).

Potential Structural Impacts

These three alternative urban economic development strategies each possess the potential to diminish the intensity of the structural dual dependencies, hence providing the necessary prerequisite for the reconstitution of urban regimes.

Lessening the External Economic Dependencies. In a general sense, all three strategies conceivably ameliorate cities' dependence on external capital by structuring a more geographically rooted local economy (see Berk & Swanstrom, 1994). Similarly, each potentially increases local economic self-reliance by building on the indigenous capacities of a city's economy to generate growth and development (see Imbroscio, 1995a). In this economic environment, public officials would find their dependence on mobile capital greatly diminished, as the city's economic vitality would be less contingent on its attraction and retention.

Consider, first, the community-based strategy. An overriding goal of this strategy is to enhance "local autonomy" in economic development —supplying individual communities with the necessary productive capacity to achieve local economic viability on their own (Bruyn, 1987). Moreover, at a narrower level, community-based economic development potentially lessens dependence on external capital in another way: Because a chief focus of this strategy is to develop the distressed areas of a city, urban fiscal pressures stemming from the existence of uneven development within cities (and the concomitant social problems) would be made less acute. As a result, the need to pursue external capital investment as a means of increasing city revenues could diminish significantly.

Likewise, the second alternative strategy—entrepreneurial mercantilism—could reduce external economic dependence. The "entrepreneurial development" (or "enterprise development," as it is often called) at the heart of this strategy conceivably provides a dynamic, diversified, innovative, and hence resilient local economy—one capable of withstanding "exogenous shocks" brought about by capital disinvestment (see Friedman, 1986; Schweke, 1985; Shapero, 1981). Harnessing the existing entrepreneurial initiative and talent in the city mitigates urban economic dependence as the local economy is, in theory, able to generate continuous economic development *from within*, lessening the need to attract outside industry. Complementing these efforts to facilitate enterprise development, the "mercantilist" development agenda of this strategy mitigates urban economic dependence by slowing the leakage of economic resources from the city and by fostering development efforts that have the greatest positive effects on the city economy.

Finally, increased municipal enterprise could be beneficial as a strategy to lessen urban economic dependence for at least two reasons. First, economic assets owned by cities would be anchored to a particular spatial location, making the issue of mobility irrelevant. Second, profits cities made from this activity would provide an alternative revenue source from which local governments could draw. As Frug (1980) writes, when the city operates profit-making ventures (as opposed to being "a receptacle solely for industries that lose money") it can "make money to serve community ends. This would not be a new idea; as late as the nineteenth century, local governments relied on profitmaking ventures to curb their dependence on taxation" (p. 1150; also see Hartog, 1983).

Lessening the Internal Resource Dependency. Earlier, I outlined the essence of the internal resource dependency faced by public officials in cities: Because of the general weakness of their authority, officials need access to extrastate resources to carry out the complex policy tasks required for the effective governance of their communities. In light of this need, the concentration of resources held by the local land-based business community makes them an attractive ally. Also recall that a reduction in this internal resource dependency can be accomplished in two ways: The resources (and hence governing capacity) of the local state can be augmented, or the distribution of resources in local civil society can be altered in a manner lessening the distinct advantage currently enjoyed by land-based business interests. As was the case for the external economic dependency, each of the three strategies for alternative economic development has the potential to reduce the internal resource dependency.

Successful community-based economic development conceivably builds a significant nonbusiness resource base in the city, as the community-based, neighborhood-oriented sector procures the fruits of economic development. Although himself skeptical, Cunningham

(1983) gives expression to the fundamental dynamic at work: Community-based economic development "could enable communal groups to build powerful, independent and prosperous new institutions" (p. 264). In this way, public officials attempting to govern their cities need not, by necessity, be drawn into close alliance with the land-based business community, because another concentration of societal resources is available. As a result, public officials' internal resource dependence on land interests would be lessened, laying the foundation for a new political arrangement.

Recent urban history provides at least one precedent for this kind of restructuring of the internal political resources in cities. Although brought about via a different mechanism, a similar phenomenon occurred during the federal government's war on poverty in the 1960s. As Peterson and Greenstone (1977, pp. 269-274) recount, the Community Action Program attacked the problem of "political poverty" of minorities in cities by funneling a variety of resources through Community Action Agencies. Armed with these new resources, minority groups were able to gain incorporation into local political systems. Although ultimately unsuccessful in breaking the governing alliance between land-based business interests and public officials, these federal efforts restructured resources in ways effecting substantial political change in cities (Elkin, 1985).

The promise of community-based economic development is that successful endeavors can produce analogous changes in the internal political resource distribution of cities, eventually lodging enough resources in community-based organizations to support an alternative governing alliance. But whereas the past efforts were ephemeral and contingent on the continued support of outside forces, community-based economic development engenders the prospect of a restructuring of internal resources that is sustainable and self-generating.[12]

The entrepreneurial-mercantilist strategy also potentially brings about the desired resource restructuring, because a major thrust of this strategy is to diversify the local private sector economy through the proliferation of small- and medium-sized businesses. Recasting the nature of the local economy through this process of "enterprise development" produces two effects that tend to reduce the internal resource dependency of local public officials. One causes the erosion of the present structure; the other eases the development of a replacement.

First, building a local economy based on the development of numerous small- and medium-sized enterprises rather than a handful of larger ones conceivably would fragment the concentration of resources currently held by the city's business community. Because much of the business community's attractiveness to local public officials stems from the existence of this concentration, fragmenting these resources could weaken significantly the current "gravitational pull" between the two groups. Second, the dispersion of individual private property ownership associated with economic change in this direction can stimulate the growth in cities of what Stone (1989) calls a "resource-rich but noncorporate middle class" (p. 288). Empirically, as alluded to above, progressive regimes often develop in cities that have an internal distribution of resources structured this way, as there is an alternative resource base available for the implementation of an alternative agenda (Stone, 1989, p. 228; Stone et al., 1991; see also Clavel, 1986). This independent, resource-rich middle class and the political organizations it creates and invigorates can be the basis of another alternative governing alliance in cities.

As DeLeon (1992a, 1992b) points out, many progressives in San Francisco understand this dynamic to be a fruitful path for building a progressive regime. As he comments, perceptively, the "political logic" of a small-business strategy and its link to progressivism is captured by Roberto Unger's (1987) notion of "petty bourgeois radicalism" (cf. Piore & Sabel, 1984, pp. 303-306): According to Unger (1987), "skilled workers and artisans, technicians and professionals, shopkeepers and even petty manufacturers" (p. 28) historically have been a force challenging established political and economic institutions (DeLeon, 1992a, p. 154). Similarly, Berk and Swanstrom (1994) point more precisely to a historical analogue of these political arrangements: the so-called producer-republican regimes of the 19th century. Such regimes were animated by "a politics of civic republicanism emphasizing equal opportunity in a commonwealth of independent producers" (pp. 27-28).

Finally, much of the same can be said of the third strategy—municipal enterprise: It, too, has the potential to reduce the internal resource dependency of public officials. The first two strategies would accomplish this goal by restructuring the resource distribution in the local civil society; this approach seeks to alter the other side of the equation—the institutions of the local state. In particular, an expansion in the munici-

pal ownership of productive assets augments the resources of local
units of state authority, creating on the local level something akin to
what Krasner (1978, pp. 55-61) conceptualizes as a "strong state"
(cf. Frug, 1980; Greer, 1987). A strong local state apparatus provides
urban public officials with an increased capacity to govern their communities effectively without the need for additional extrastate resources. Hence their current internal resource dependency on the local
business community would be diminished significantly.

EVALUATING THE STRATEGIES

Thus, our three strategies for alternative local economic development
—community-based economic development, entrepreneurial mercantilism, and municipal enterprise—have the potential to lessen the structural dual dependencies of urban public officials. Following the
argument outlined above, this, in turn, would facilitate a reconstitution
of current urban regimes in ways promoting our ultimate aim: an
increase in the normatively desired (liberal) political equality in contemporary central cities.

The current development of these strategies in cities is largely sporadic and embryonic—they are basically experimental in nature. As a
result, we cannot, at present, fully evaluate their ability to achieve the
above goals. Nevertheless, if there is at least a possibility that any of
these strategies can achieve these goals, at a minimum a strategy must
meet two conditions: A strategy must be (a) *effective* in bringing about
urban economic development and vitality and (b) *feasible* to implement
in cities on a scope allowing it to have a significant impact on the
structural context of urban regimes.

The first condition—the effectiveness issue—requires assessing the
prospects that a strategy "works" in the broad sense of doing what a
development strategy is supposed to do—for example, generate net
new employment, increase aggregate personal income, expand the tax
base, and so on. The second—the feasibility issue—entails identifying
the existing impediments to a strategy's fuller development and gaining an understanding of whether such impediments can be overcome.

In the following chapters, I evaluate the three strategies by exploring
their potential to meet these conditions. Because these strategies are

TABLE 2.1 The Potential of the Alternative Strategies to Lessen the Structural Dual Dependencies

	Alternative Strategies for Urban Economic Development		
Structural Features of the Urban Polity	*Entrepreneurial Mercantilism*	*Community-Based Economic Development*	*Municipal Enterprise*
Reducing external economic dependence	Indigenous entrepreneurial renewal Slows economic leakage/Maximizes local enefits	"Local autonomy" in development Lessened uneven development	Anchored enterprise Public profits
Reducing internal resource dependence	Fragments business resources Stimulates resource-rich, noncorporate middle class	Builds a community-oriented resource base	Builds a strong local state

highly experimental in nature and represent a fundamental shift from the current practice of urban economic development, I begin my exploration of each strategy's prospects for effectiveness with the assumption that they will *not* be effective in bringing about economic development; I then marshal a variety of empirical evidence pointing to the opposite conclusion. To explore the feasibility question, I conduct case analyses, where productive, of cities experimenting with individual strategies. Before undertaking these two tasks, however, in each of the following chapters I first attempt to provide a better conceptual understanding of each strategy by exploring its philosophical vision and substantive core.

NOTES

1. The debate's designation derives from the controversy surrounding Paul Peterson's (1981) seminal work of the same title. Also see Henig (1992) and Sonenshein (1988).

2. See Swanstrom (1988, p. 107), who also uses the distinction between "the external economic structure" and the "internal political dynamics" of the urban political economy. In addition, see Kantor (1988) who, in a slightly different way, also uses this helpful analytical distinction between external and internal factors in urban political economy.

3. Most prominent here are the works of Marxian and neo-Marxian urban scholars, especially those influenced by structuralism (see Harvey, 1973). Kantor (1988) also focuses heavily on external economic factors while applying an essentially Lindblomian framework (see Lindblom, 1977). Gottdiener (1987) offers another analysis in this vein, relying on the insights of state theory written from both a neo-Marxian and a neo-Weberian perspective.

4. Compare DiGaetano (1989) who also assesses the literature on urban regimes and makes a similar observation concerning this internal perspective: "This school of thought," he writes, "often relegates the market, as a structuring force, to a minor or unspecified role in regime formation" (p. 263).

5. Peterson (1981), for example, argues that economic handcuffs compel cities to pursue pro-business development policies and avoid those policies with a redistributive intent. Given the nature of the environment cities face, Peterson deems this policy agenda to be objectively rational and, as such, promotes the city's interest as a whole.

6. What Stone (1986) points out regarding Atlanta applies to most other major cities. He notes that although these enterprises can vary their commitment to invest in the city, "banks and other large businesses have sunk investments in the central business district, and they cannot simply walk away from these investments. Major holders of commercial property cannot afford loose talk about disinvestment. If such talk got out of hand, it might engender a self-fulfilling prophecy." (p. 99)

7. In a manual designed to instruct local public officials on how to strengthen their economies, the Committee for Economic Development (1982), a research organization of elite business executives, points to the importance of a community's business climate and makes recommendations on how local governments can make the necessary improvements. Among other probusiness policies, a key element of the organization's formula

"involves extensive efforts by local government to communicate with the private sector" (p. 31).

8. Of course, the two overlap. Both stem from "the division of labor between market and state as [it] . . . is manifest in cities" (Elkin, 1987, p. 18). Separating the two dynamics, however, will allow us to better delineate the *consequences* of that general formulation. Moreover, the effects of each dependency are relatively independent of the other, as neither can be reduced to the other. For example, the forces behind this second dependency would be at work even in absence of any significant external economic pressures on a city (cf. Stone, 1991, pp. 292-293).

9. Two major factors—one economic, the other organizational—help explain the presence of this concentration. First, the nature of the business corporation itself produces a situation in which relatively few actors have control over a massive amount of resources (Stone, 1989, p. 229). And second, as a group, downtown (land-based) business leaders are often well organized into strong peak associations that articulate and attempt to implement a reasonably unified political program.

10. The alliance can also be empowering for land-based business—making the internal resource dependency a *mutual* affair. Although the authority of the local state is circumscribed, public officials do control some resources beyond the reach of private actors. Perhaps most crucial is that the use of public authority in democratic societies usually carries with it a strong measure of legitimacy. As a result, by working with public officials these interests can accomplish more of their agenda than would otherwise be the case.

11. Note that each of these reforms of the second dependency—the internal resource dependency—aspire to reconstruct different institutional realms in the urban community—the first, the (local) state; the second, civil society.

12. Compare the work of Hunter (1980), who advocates a version of the community-based paradigm as a means of altering the undemocratic governing arrangements he finds in Atlanta.

PART II

Analysis

THE ENTREPRENEURIAL-
MERCANTILIST STRATEGY

This chapter examines the first of the three alternative strategies for urban economic development—what I have labeled "entrepreneurial mercantilism."[1] A sketch of its overall vision and substantive core sets the context for the later analysis of its effectiveness and feasibility.

EXPLORING VISION AND SUBSTANCE

Vision

First and foremost, the vision projected by the "entrepreneurial-mercantilist" strategy for local economic development owes much to Jane Jacobs's (1961, 1969, 1984) writings on the dynamics of urban capitalism. For Jacobs, urban economic vitality results when a physical, social, and financial climate supportive of small-scale entrepreneurs is created and sustained in a city. Such an environment promotes what she calls the "two master processes" of economic life: innovation and import replacement (Jacobs, 1984, p. 39). Import replacement, which she sees as the source of urban economic expansion, occurs as entrepreneurs begin to produce goods and services locally that were once imported. This process is itself made possible through the development of economic innovation. Economic innovation, as Polsky (1988) notes in his analysis of Jacobs's writings, "occurs when entrepreneurs learn to add new forms of work to established types of labor" (p. 10).[2] This in turn requires creativity, and small-scale businesses—being more flexible than large, bureaucratic corporations—are more likely to be the source of creative economic behavior (see Schweke, 1985).

Entrepreneurial mercantilism envisions the development of an urban economy modeled on the Jacobs ideal. Such an economy, based on a variety of small entrepreneurial businesses, would be highly diversified. Because of this diversity and the capacity for creativity and innovation, the economic process at work in this idealized setting, although messy and inefficient, would be highly dynamic and resilient in the face of technological and economic change—and thus capable of achieving development that is both self-sustaining and self-renewing (Jacobs, 1969; Malizia, 1985, pp. 40-41; Shapero, 1985, pp. 210-211).[3] Following her conceptions further, this strategy imagines a highly interdependent local economy in which, as one author notes, "a complex network of economic interrelationships . . . in which a web of producers and suppliers, [sic] secondary, tertiary and further producers and suppliers, wholesalers, retailers, and consumers are joined" (Meehan, 1987, p. 136). A healthy economy, argues Jacobs, is one where economic actors exist in a kind of symbiotic relationship with one another, as each actor fills particular niches in the system in ways that strengthen the local economic process as a whole (Jacobs, 1984; Meehan, 1987, p. 144).

Second, the vision of the entrepreneurial-mercantilist strategy is also animated by the ecologist's dream of constructing a society based on the principle of "local self-reliance." The key idea here is to reorient the process of economic production in ways that *use local resource bases as the chief means of satisfying local needs* (Bookchin, 1982; Sale, 1980; Schumacher, 1975, p. 55). This reorientation provides for ecologically intelligent forms of economic development, as an economy structured in this "radically decentralized" manner forces localities and their citizens to "pay great attention to the life-support capacities of the ecosystem(s) upon which they rely" (Dryzek, 1987, p. 218). Achieving "local self-reliance" requires, most crucially, the creation of a number of so-called closed ecological loop systems in which the wastes of one production process are transformed into the raw materials of another (Meehan, 1987; Morris, 1982a, p. 5). More generally, it requires closing the gap between consumer and producer. Such processes enhance "local self-reliance" by limiting the dependency of local communities on outside resources.[4]

Another aspect of this strategy's vision entails conceptually reviving the older image of the city as a "walled" entity. Unlike the Greek city-state and the medieval and Renaissance city, an "unwalled city," writes the venerable Norton Long (1972, pp. 3-4), has no "significant, real boundaries." More important, its economy is completely open, and hence the city is extremely "vulnerable to buffeting by social and economic forces beyond its control" (Judd & Ready, 1986, p. 209). In contrast, the entrepreneurial-mercantilist strategy for local economic development takes seriously the idea of the city as an independent economic system. So, for example, David Morris (1982a), a leading theorist of this perspective, notes that in this approach a city conceives of itself "as a nation," analyzing the movement of capital within its borders and evaluating its "balance of payments" (p. 6). Furthermore, in its redefinition of the city's development strategy, entrepreneurial mercantilism envisions that the integrity of its "walls" can be (at least partially) restored. This restoration is accomplished as the city economy is restructured in ways making it relatively immune to the effects of exogenous forces.

Finally, underlying all of this is a renewed interest in the virtues and power of localism, as well as an emphasis on building community and a strong sense of place. Rounding out this vision, we find—at its heart —an essentially inward-looking, spatially biased "mercantilist" philosophy: The explicit goal of the city's intervention into the local economic development process is to maximize the benefits accruing to the *local* economy from that process.

Substance

What constitutes the core of the substance of the entrepreneurial-mercantilist strategy for alternative local economic development? That is, how would a city act when it has chosen to implement this strategy? To answer this question in the broadest way, we can identify two fundamental principles that would serve as the guide for all of a city's economic development efforts. The first involves the pursuit of *indigenous* economic development: Rather than being based on the attraction of outside economic resources, the emphasis here would be on generating local development and vitality from within the local economy itself (cf. Eisinger, 1988). A related, second principle involves a strong focus on a *city's resource flows*. Along these lines, a concern for slowing the "leakage" of economic resources from the city would be pronounced, as would be an attempt to foster those development efforts having the greatest positive effects on the *city* economy (Morris, 1982a, p. 6; see also Giloth, 1988, p. 348).

From these two general guiding principles, seven specific goals of entrepreneurial mercantilism follow. As will become apparent, many of these goals mutually reinforce one another, and together work to bring about the more general outcomes desired by cities following this economic development path. In what follows, I outline these seven goals individually, including for each (a) a brief description, (b) a discussion of its salutary contribution to the local development process from the perspective of entrepreneurial mercantilism, and (c) a brief note on the general means a city would employ to realize each goal.

Goal 1: Stimulating Small Business ("Enterprise Development"). The first goal of this strategy is to stimulate small-scale enterprises operating in the city by aggressively assisting and supporting the efforts of indigenous entrepreneurs. Viewed through the lens of entrepreneurial mercantilism, this activity greatly benefits the local development process for at least three reasons. First, drawing on the theoretical work of Jane Jacobs discussed above, a healthy, entrepreneurial small-business sector promises to contribute to the creation of a dynamic, diversified, innovative, and hence resilient local economy—one capable of withstanding exogenous shocks brought about by capital disinvestment (see Schweke, 1985; Shapero, 1981). Second, because small businesses need a variety of support services, they are likely to have substantial linkages to the local economy, further enhancing local economic development through their spin-off purchasing practices (Fisher, 1988, p. 175). Finally, these small enterprises are the type of companies more likely to

TABLE 3.1 Goals of Entrepreneurial Mercantilism

Stimulating Small Business (enterprise development)

Description:	The rapid formation/expansion of small businesses
Benefits:	Resilient local economy, spin-off purchasing, jobs for city residents
Means:	Financial/technical assistance, including help locating new markets, seed capital, incubators

Promoting Local Ownership

Description:	Independent ownership of business enterprise/other productive assets
Benefits:	Spin-off purchasing, profit recirculation, less mobility, employment benefits
Means:	Same as efforts to stimulate small business, also assisting local buyers of businesses

Increasing Import Substitution

Description:	Reliance on imports decreased via the local production of goods/ services now imported
Benefits:	Reducing leakages, greater local self-sufficiency, enhanced resource productivity
Means:	Identifying imports to replace/assisting local business, buy local campaigns, city procurement, use of natural resources

Conserving Resources

Description:	Limiting city's consumption via increased efficiency of resource usage
Benefits:	Less dependent on outside forces
Means:	Energy efficiency of built environment, transportation, etc.; aggressive recycling programs

Strengthening Local Multipliers

Description:	Enhancing the economic activity generated in the local economy, more interindustry dependence
Benefits:	Increased local economic self-sufficiency
Means:	Same as small business development, local ownership, and import substitution; also manufacturing/other high multiplier activities

Tapping Innovative Local Finance

Description:	Locally generated financial resources stimulate the local economy to the maximum degree
Benefits:	Less need to attract outside financial capital for economic development
Means:	Municipal pension funds, linked deposits, mini-munis, local currencies

Localizing Employment Policy

Description:	Maximize the number of city dwellers obtaining jobs in the city economy
Benefits:	Slows leakage of economic resources from city
Means:	Residency requirement for municipal employees, mandating firms to hire city residents

hire a larger proportion of *city* residents, thus keeping the benefits of employment in the city economy from escaping to the suburbs.[5]

Economic development officials in cities pursuing this element of the entrepreneurial-mercantilist strategy would provide a range of needed financial and technical assistance to small local firms, perhaps through the creation of a small-business assistance center (Blakely, 1989, pp. 164-165).[6] Of particular importance would be helping these enterprises locate and exploit new markets for their goods and services, including demonstrating to them how they can best take advantage of opportunities to market their products to the public sector (Schweke, 1985, p. 5), as well as aiding their efforts to trade with overseas companies (Eisinger, 1988, chap. 12). Moreover, given the short life expectancies of many small businesses, and because many successful, mature small companies may leave the city that nurtured them in their infancy (Fisher, 1988; Jacobs, 1984), a crucial element of any local economic development strategy focused on stimulating small business would involve fostering the rapid generation and expansion of *new* enterprise in the city (see Ahlbrandt & DeAngelis, 1987; Bergman, 1986, p. 10; Friedman, 1986; Schweke, 1985).

To generate new "enterprise development," cities would provide "seed capital" for start-ups and expansions of businesses unable to easily obtain financing from conventional sources (see Schweke, 1985). This seed money could take the form of high-risk venture capital for companies with considerable growth potential;[7] on the other end of the financing spectrum, cities might offer "micro loans" (of approximately $5,000) to individual entrepreneurs wishing to start home-based microbusinesses.[8] Beyond this, cities would establish small-business "incubators"—"greenhouses" for cultivating new, developing companies —in which a city provides low rent industrial space, a set of business services, and management training to a number of companies sharing a building (see MacDonald, 1985; Plosila & Allen, 1985).[9]

Goal 2: Promoting Local Ownership. Another key goal of this strategy is to encourage local (i.e., independent) ownership and control of enterprises and other productive assets. This economic development effort advances the objectives of entrepreneurial mercantilism in numerous ways. Like small businesses, locally owned businesses generate spin-off purchasing in the local community: As one analyst notes, "absentee-owned businesses . . . are less likely to purchase local services and products, such as legal assistance, financial consulting, capital borrowing, and factor inputs" (Morris, 1982a, p. 38). More generally, local ownership contributes to a high rate of profit recirculation in the local economy, as "profits are not siphoned off to corporate headquarters in other cities" (Fisher, 1988, p. 175). Local ownership also increases the likelihood that assets will remain in the local area, slowing the pace of the mobility of capital (Alperovitz & Faux, 1984, p. 146; Fisher, 1988, p. 175). Moreover, locally controlled businesses increase employment at a faster rate than absentee-controlled companies (Schweke, 1985), and during an economic downturn these companies tend not to cut employment levels as much as their nonlocal counterparts (Alperovitz & Faux, 1984, p. 146; Bluestone & Harrison, 1982, p. 164).

To achieve local ownership, the efforts undertaken by city economic development officials to stimulate small businesses will also be important here, as most small enterprises will be locally owned. Beyond this, officials would promote local ownership of the city's economic assets by, for example, locating and assisting local buyers for businesses available for purchase.

Goal 3: Increasing Import Substitution. A third objective of this strategy is to decrease a city's reliance on imports. The aim is to stimulate the local production of goods and services—both producer and consumer —currently imported. Discussing the advantages of this economic development practice, Alfred Watkins (1980) explains that a city substituting for imports "enhances local prosperity by reducing leakages [of funds from the city, increasing the amount of money circulating through the local economy]. More jobs are created within the urban economy, there is a greater degree of local self-sufficiency, and less funds escape to a rival economy" (p. 133). Moreover, as Persky, Ranney, and Wiewel (1993) add, "a program of import substitution mobilizes local resources in a fashion that greatly enhances their productivity. . . . [It] may be a socially useful way in which to offset what would otherwise be the waste of redundant resources" (p. 121).

The most obvious means to this end would find a city's development officials identifying the types of imported products that can be replaced most fruitfully by local production, informing local companies of these opportunities, and encouraging them to compete for this business

(Ahlbrandt & DeAngelis, 1987, pp. 47-48), while at the same time providing the necessary technical and marketing assistance. The flip side of these efforts would have officials launching a "buy local" campaign, using the mass media and other communication strategies to persuade local businesses and residents to purchase goods and services from hometown suppliers (see Meyer, 1991, p. 173).[10] Moreover, by increasing its *own* purchasing from local companies, the city government itself can replace imports through its procurement practices.[11] Finally, an even more innovative method of import replacement results when local businesses and individuals use previously unused raw materials and natural resources found in the city's environment to produce goods locally.[12] These raw materials might include, for example, locally generated waste products that are recycled and substituted for imported materials. More generally, cities tapping into their natural potentials would focus heavily on encouraging efforts that increase local production of two of their most expensive imports: food and energy (see Benello, 1988; Morris, 1982a, pp. 51-60; 1982b).

Goal 4: Conserving Resources. Another aspiration of the entrepreneurial-mercantilist strategy is to promote resource *conservation.* Although closely related to the local resource utilization efforts discussed above, this goal is conceptually distinct: Rather than trying to tap unused local resources to meet local production needs as a means of increasing the city's self-reliance, the emphasis here is on the more straightforward task of making the city more self-reliant by limiting its overall consumption of resources. As the city as a whole consumes less, the logic of this measure asserts, its total dependence on outside forces beyond its control will diminish. The key is getting the maximum output from the minimal input—that is, increasing the efficiency of resource usage. Cities pursuing this goal would, for example, work to increase the energy efficiency of their built environments, transportation systems, and industries. They might also start aggressive recycling programs to improve local resource-use efficiency by limiting the need for costly municipal solid waste disposal (see Morris, 1982b).[13]

Goal 5: Strengthening Local Economic Multipliers. Local multipliers quantitatively represent the amount of economic activity ultimately generated in the local economy as a result of some initial local economic activity. Increasing these figures is the fifth goal of entrepreneurial mercantilism. With high economic multipliers in a local economy, any small degree of local economic activity stimulates a great deal more economic activity in the local economy. As a result, an urban economy with high internal multipliers enhances the city's economic self-sufficiency.

Many of the economic development goals of the entrepreneurial-mercantilist strategy already explicated above—small-business development, local ownership, import substitution—would conceivably augment local multipliers. This is because each of these undertakings increases the level of *interindustry dependence* in the city economy, where the purchasing and procurement activities of any one local industry stimulates other local industries. Thus, by working toward those three goals, local development officials would go a long way toward increasing economic multipliers in their cities. Moreover, in sectoral terms, research indicates that the multipliers generated by manufacturing industries usually outstrip those generated by industries in other sectors of the economy (see Bluestone & Harrison, 1982; Carlson & Wiewel, 1991, p. 19).[14] As a result, efforts to encourage manufacturing would also be used to achieve this goal.[15] Finally, at a more general level, officials would evaluate new businesses locating in the city by the degree to which these businesses would increase local economic multipliers and, in light of this evaluation, would either encourage or discourage such development.[16]

Goal 6: Tapping Innovative Local Finance Sources. Much of the financing for this strategy would need to be obtained through the reorientation of the currently substantial expenditures made by cities on corporate-center/mainstream economic development efforts. Nevertheless, this approach also comes equipped with its "own" financial mechanisms, innovative financing schemes consistent with the guiding principles and logic of entrepreneurial mercantilism. Such schemes attempt to alter local financial arrangements in ways that would allow locally generated resources to stimulate the local economy to the maximum degree.

At least four different innovative finance practices meet this criterion. First, cities would invest portions of their municipal pension funds in local economic development, so that economic resources that otherwise would escape from the city can be captured (Keating, 1991, p. 167;

see also Leatherwood, 1983; Rifkin & Barber, 1978, chap. 18).[17] Second, cities would establish a "linked deposit" program, depositing its temporarily idle revenues (which accumulate just after tax collections or bond sales) only in private banks agreeing to make investments back into the local economy (see Litvak & Daniels, 1979, pp. 118-124; Rosen, 1988, pp. 121-123).18 Third, cities would issue (and use mass-marketing techniques to advertise) tax-free "minibonds" or "minimunis" —smaller-denominated securities (in amounts as small as $250) that allow moderate-income residents to invest their personal savings in their municipalities and shelter their income from centralized taxation at the same time.[19] Finally, cities might create their own local currencies to supplement the conventional cash flow in their jurisdictions.

Though the first three of these financial schemes are relatively straightforward, the fourth requires further explanation. To establish a local currency, a community would set up and administer a number of accounts (similar to bank accounts) for local residents and local businesses joining the system. Local members could then trade or "barter" goods and services with one another, and these transactions would be recorded in the relevant individual accounts. "Money is just a form of social information," asserts a brochure publicizing this concept: A local economy often "suffers from unemployment of its people or other resources such as land, equipment, or energy, simply because there is not enough money circulating in the community"; hence it "can bring these resources into effect by organizing a local currency" (Landsman Community Services, 1989).[20] Most crucially, this measure ensures that local commerce would bring significant local benefits, as those wishing to use their account balances to purchase items would need to contract with other members of the local currency system.

Goal 7: "Localizing" Employment Policy. Finally, the logic of the entrepreneurial-mercantilist strategy implies that the city must make efforts to "localize" its employment policy. This "localization," as understood here, means that the city would try to maximize the number of city dwellers obtaining jobs in the city economy (both public and private sectors). Such actions are desirable because they slow the leakage of economic resources from a city's boundaries: Out-migration by the urban citizenry is discouraged, city unemployment is lowered, and the salaries earned by urban job holders circulate more in the local economy (Eisinger, 1983, p. 94; Judd & Ready, 1986, pp. 245-246).

As mentioned earlier, encouraging the development of certain types of businesses in the city—namely, "small" companies—can work to maximize the number of city residents employed in the city (Long, 1987). In contrast, this seventh goal involves more direct measures to achieve this objective. Two of the most common measures used by cities are (a) to require municipal government employees to be residents of the city and (b) to require developers and construction firms operating in the city to hire city residents.[21]

ASSESSING PROSPECTS FOR EFFECTIVENESS

Chapters 1 and 2 advanced the following propositions: To bring about the reconstitution of urban regimes in a manner that enhances the degree of political equality in central cities, it is necessary to diminish the structural dual dependencies faced by urban public officials. The three alternative strategies have the potential to bring about this diminution. If this potential is to be realized, however, the alternatives must be *effective* as economic development strategies.

Below, I explore the effectiveness question for entrepreneurial mercantilism, examining both overarching and more specific issues. For this exercise, as noted in Chapter 2, the initial assumption is that these strategies will not be effective, given that they represent a fundamental shift from current urban economic development practice. In response, I marshal a variety of empirical evidence to the contrary.

Overarching Issues

Two important overarching issues are the strategy's relationship to orthodox economic theory and to established legal doctrine. As "alternative" approaches to urban economic development, the three strategies challenge, to varying degrees, the prevailing economic orthodoxy —neoclassical theory (cf. Albelda, Gunn, & Waller, 1987, p. 3). If orthodox economic theory is indeed correct about these disputed matters, that would limit significantly a strategy's likely effectiveness as a development strategy. Likewise, if the courts are likely to disallow key elements of a strategy, its prospects for effectiveness are similarly reduced.

Economic Theory

In the case of entrepreneurial mercantilism, the challenge to the prevailing economic orthodoxy can be traced to its spatially biased 'mercantilist' demeanor. The economic philosophy of mercantilism—including municipal mercantilism—was, after all, thoroughly attacked more than 200 years ago in Adam Smith's *The Wealth of Nations*, the work that first systematically developed the key premises of classical economics (Judd & Ready, 1986, p. 237). In a more specific sense, the spatial biases of this strategy point to two theoretical deviations from the economic orthodoxy.

First, to generate urban economic growth, this strategy strongly (but not exclusively) emphasizes strategies to increase production efforts targeted to *meet local needs* (both consumer and producer). This central notion is at the basis of the development goals of increasing import substitution, strengthening interindustry dependencies and, to a lesser extent, stimulating small-business development and promoting local ownership. This, however, makes the entrepreneurial-mercantilist approach inconsistent with a key theoretical tenet of mainstream urban-regional economics—that the vitality of a city's economy is most critically determined by the expansion rate of those industries that produce goods and services for an *external* (nonlocal) market (cf. Thompson, 1965; see also Kantor, 1988, pp. 14-15). These "export" industries are seen as particularly valuable to the local economy because (a) they draw new money into the community from outside and (b) to function, they usually require an array of supportive businesses (Sharp, 1990, p. 222). Echoing a long tradition of urban-regional economists, Peterson (1981, p. 23) concludes that because these twin benefits do not ensue with the production of products for local consumption, local citizens "are simply taking in one another's laundry."

Though some exporting is no doubt always necessary for a local economy to thrive, some evidence suggests that this "export-based" theory—with its obsessive focus on the capture of external markets as the sole path to local prosperity—is itself flawed (Persky et al., 1993, pp. 19-20; Fusfeld & Bates, 1984, p. 141). "If growth in per capita income occurs only through expanded sales outside the region," Vaughan (1988) points out, "it is difficult to explain the increase in global income without reference to intergalactic trade" (p. 120). Moreover, in direct challenge to the dominant tradition in urban-regional economics, Vaughan continues by noting that

> growth can come as easily from within as from without. A new local laundry that cleaned more effectively and at a lower cost improves our well-being, which is, after all, the ultimate goal of development. It also leaves us more to spend on other goods and services—some of which will be locally produced. And the successful proprietor may serve as a role model for, and investor in, local entrepreneurial ventures—vital development functions that are conveniently omitted from most growth models. (p. 120)

In fact, an overreliance on the advice of economists working from this "grossly oversimplified" model has led public officials at both the state and local levels to make imprudent economic development policy decisions. In particular, Vaughan (1988) cites the "chronic undervaluation of the importance of human capital, innovation, and entrepreneurship in development" (p. 120).

Norton Long (1987) reaches a similar conclusion: "Cities," he writes, "have generally been accustomed to targeting the export sector because of [orthodox] economists' emphasis on the importance of the export base for the local economy" (p. 197). Although the importance of exporting for the local economic vitality is undeniable, "not all export industries are accompanied by a stimulation of the rest of the local economy, and some seem to have little favorable effect upon it or may even be perverse in this respect" (p. 197). For those export industries most important in contemporary cities—the downtown advanced service sector—this proposition seems to hold: That sector is generally isolated from the "local" component of the urban economy because the economic activities of the advanced service companies often lack a strong link to a city's small- and medium-size local businesses (Stanback & Noyelle, 1982, pp. 140-142).

Examining entrepreneurial mercantilism's more general preoccupation with maximizing *local* benefits from the local development process—and, in particular, its emphasis on stimulating *indigenous* economic development—reveals a second challenge to a related, yet even more venerable tradition in mainstream economic theory: the doctrine of free trade. Following the "law of comparative advantage" and the corresponding "gains from trade" argument first set out by classical economists, orthodox economic theorists deduce the following: To promote the economic well-being of a geographical area,[22] public policies ought to create conditions promoting the purchase of production factors and finished goods and services from outside the local economy, if these items can be produced cheaper abroad. The substance of entrepreneurial mercantilism, in contrast, attempts insofar as possible to promote *local* production and resource usage as a means of meeting local needs.[23] Therefore, in theory at least, this strategy imposes economic costs and efficiency losses on the local economic system and, as such, severely reduces the prospects that it can bring about local economic vitality.

This orthodox theoretical position can be attacked, however, as being one-sided—as examining only one side of the equation: Economic *advantages* as well as economic costs can accrue from this strategy. These advantages arise for three reasons. First, not using more fully available local economic resources entails substantial opportunity costs (Persky et al., 1993, pp. 19-21); second, the local economy is more stable, being less vulnerable to the vicissitudes of external economic change (cf. Morris, 1982a); and third, local actors retain more control over their economic destiny (cf. Meehan, 1987). In short, the costs imposed on the local economy perhaps can be ultimately balanced by the resulting economic advantages to the local economy.

Hence, the entrepreneurial-mercantilist strategy's deviation from orthodox economic thought on these two counts does not necessarily detract significantly from its prospects for effectiveness. Clearly, we must move beyond the realm of economic theory to gain a more adequate understanding of the prospects for this strategy's effectiveness.

Legal Context

The spatially biased, "mercantilist" demeanor of this strategy not only raises questions regarding the soundness of its economic theory; it also necessitates that we scrutinize this approach from a jurisprudential angle. In fact, the constitutional environment in the United States severely proscribes the *means* available to cities wishing to follow this economic development path. As David Morris (1982a) writes by pursuing these objectives,

cities emphasize spatial considerations that undermine one of the principle tenets of our Constitution—the continental free-trade zone. The
unencumbered mobility of goods, capital, and people across state and
city boundaries was for many of the Founding Fathers the chief purpose
of the Constitution. (p. 8)

A city cannot, for example, help local producers substitute goods for
those imported into the city by using its regulatory powers to prohibit
nonlocal goods from entering the city's marketplace. Such a "protec-
tionist measure" would clearly be in violation of the Commerce Clause
of the U.S. Constitution, which prohibits states or localities from burdening the flow of interstate commerce.

Yet none of the means to realize the goals of entrepreneurial mercantilism outlined above violate the constitutional provisions of the Commerce Clause. Historically, the means coming closest to being declared
violative of this clause have involved attempts by a city to maximize
economic development benefits from its procurement and employment
policies. City actions that discriminate in favor of local suppliers for
governmental purchasing and require private companies receiving city
funds to employ a percentage of local residents have both received
serious scrutiny. The judicial system, however, has deemed these policies to not be in violation of the Commerce Clause on the grounds that,
when local governments engage in these actions, they do so not as
"market regulators," but as actual "market participants." The argument
accepted by the Supreme Court is that "state and local governments
have a measure of immunity . . . under the Commerce Clause when
contracting with private parties to buy or supply goods and services"
(Clark, 1985, p. 105).[24]

In light of the intent of the framers of the Constitution, it is not
surprising to discover that the Commerce Clause is not the only provision serving as the basis from which to impugn the constitutionality of
the spatially biased local actions of entrepreneurial mercantilism. For
example, attempts by local governments to obligate private companies
receiving public funds to hire city residents also have been challenged
as violations of the "Privileges and Immunities Clause" of the Constitution. Courts generally have looked more favorably on these challenges, as a city's role as a "participant" in the marketplace affords them
less protection from the constitutional restrictions in this clause
(Mandelker et al., 1990, pp. 396-397).[25] Nevertheless, the Privileges and

Immunities Clause is not seen as an absolute by the judicial system, and if municipal actions can be shown to be justified by the gravity of the economic and social ills stemming from urban unemployment, they will usually withstand constitutional challenge (Mandelker, 1990, p. 397; McCarthy, 1990, p. 248).[26]

Another constitutional provision often invoked in cases contesting elements of the entrepreneurial-mercantilist strategy is the Equal Protection Clause. In particular, the Equal Protection Clause has been a key basis for challenging the constitutionality of those laws mandating that local government workers must live in the city. These "municipal employee residency requirements" allegedly discriminate against non-residents by denying them their constitutional right to travel freely (Greene & Moulton, 1986, p. 186). Courts, however, have rejected this argument, and have held that these residency requirements do not violate the federal Constitution as long as they are "reasonable, i.e., where there is a reasonable link between the residency and the position for which it is required" (McCarthy, 1990, p. 102).[27]

Specific Issues

We now turn to more specific issues relevant for assessing entrepreneurial mercantilism's prospects for effectiveness. I divide this discussion into three general topics: (a) small-business development, (b) other economic development objectives, and (c) finance.

Small-Business Development

To a substantial degree, the prospects for effectiveness of the entrepreneurial-mercantilist strategy are linked to the proposition that successful small-business development can be the basis of sustained economic vitality in the city. Not only is the stimulation of these enterprises a key entrepreneurial-mercantilist goal itself, this action is also an important means of achieving many of this strategy's other developmental goals, such as promoting local ownership and strengthening local economic multipliers.

One source of empirical evidence supporting the conjecture that local economic vitality can stem from successful small-business development is the body of research indicating that small businesses, despite their higher failure rates, generate a disproportionate number of new jobs through the rapidity with which they organize and grow. The earliest and most influential study advancing this claim was David Birch's (1979) *The Job Generation Process*. On the basis of an analysis of data collected by Dun & Bradstreet on 5.6 million businesses, he showed that between 1969 and 1976, businesses employing fewer than 20 employees generated two thirds of all net new jobs, and those employing fewer than 100 employees generated 82% of all net new jobs. Moreover, Birch's findings also pointed to the importance of *new* enterprise development as a major source of job creation, as he discovered that businesses younger than 5 years old created about 80% of the replacement jobs during this period.

Subsequent studies challenged Birch's original findings, however. Most significantly, Brookings Institution researchers Armington and Odle (1982) criticized Birch's use of "establishments" as his unit of analysis.[28] When these researchers instead used the "firm" as the unit of analysis, the same Dun & Bradstreet data file used by Birch showed that the percentage of net new jobs generated by "small businesses" dropped significantly.[29] Nevertheless, the Brookings studies confirmed a key tenet of Birch's analysis (Eisinger, 1988, p. 239): Those small, independent companies with fewer than 100 employees, though accounting for only 36% of all employment in 1976, actually generated 51% of net job growth between 1976 and 1980 (Armington, 1983).[30]

Moreover, the quality of the Dun & Bradstreet data used by Birch and others has come under attack (Vaughan & Pollard, 1986; White & Osterman, 1991).[31] Yet other studies of job generation based on the more comprehensive data gleaned from state unemployment compensation records (ES-202 files) also point to the disproportionate influence of small businesses as job creators, corroborating the findings of the studies using Dun & Bradstreet data files.[32]

To conclude, then, we can summarize by invoking the succinct comments of one observer of this controversy: "Although methodological disputes exist, all studies agree that small companies create more jobs than their share of total employment" (Allen, 1989, p. 298; see also Zipp, 1991).

Studies of job generation are underlain in part by a second factor also tending to favor an economic development strategy that emphasizes small-business development: the recent structural transformations in the organization of economic production brought about by heightened international competition. As a response to new competitive conditions, there is a "relative decline in the importance of Fordist mass production and an enormous expansion of manufacturing activities based on less rigid and more highly adaptable (i.e., 'flexible') technological and institutional structures" (Scott, 1988, p. 171; see also Harvey, 1991; Mayer, 1988; Sabel, 1982). The ascent of this "flexible accumulation" or "flexible specialization" as the hallmark of economic production in the "post-Fordist" era magnifies the consequence of smaller companies in the economic development process (Noyelle, 1986, pp. 12-13). Such companies, for example, often embody the necessary flexibility in their production and management systems to respond adroitly to the kinds of new competitive challenges for rapid product innovation that led to the relative decline of mass production (Eisinger, 1988, p. 241; see also Piore & Sabel, 1984).

Smaller businesses are of increasing importance in the new post-Fordist era for another reason—namely, the replacement of the principle of vertical integration by the principle of vertical disintegration as a key organizational trait of production companies. Whereas vertically integrated companies establish their economic clout by controlling a larger scope of production activities internally, vertically disintegrated companies enhance their flexibility by relying extensively on other companies to supply the products and corporate services necessary for production.

This externalization (or outsourcing) of production augments the role played by smaller businesses in the economy in two ways. First, because of the downsizing accompanying vertical disintegration, individual companies themselves are more likely to be smaller establishments.[33] Second, and more significantly, the externalization of a company's production activities creates fertile market opportunities for numerous other (often small) companies (Noyelle, 1986, pp. 12-13; Scott, 1988, p. 174). The latter phenomenon can be viewed as especially propitious for *local* economic development, as, in spatial terms, it tends to lead to the formation of agglomeration economies where "flexible production networks" of market-interdependent manufacturers cluster together to make subcontracting easier (cf. Hatch, 1991; Noyelle, 1986, p. 13; Scott, 1988, pp. 178-179).[34]

On the other hand, despite these advantages, a potentially serious drawback hampers the use of small-business development as a key component of an urban economic development strategy. This drawback centers on the question of *job quality:* Specifically, these businesses may be prolific employment generators quantitatively, but how much do these jobs contribute to a city's overall level of development? Many scholars claim that small businesses usually pay low wages, offer only part-time or unstable jobs, and practice overt employment discrimination. Otherwise put, employment options in many small companies tend to be a "secondary labor market" (Bluestone & Harrison, 1982, pp. 221-222; cf. Brown, Hamilton, & Medoff, 1990; Preteceille, 1990, pp. 43-44).

Nevertheless, these claims have not been empirically sustained. For example, individual jobs with secondary labor market characteristics are, on average, found in business establishments of all sizes (Zipp, 1991, pp. 19-20). In short, the evidence demonstrating that "employment expansion in new and small business creates lower quality jobs overall" remains inconclusive (Vaughan & Pollard, 1986, p. 126). Thus, although the issue of job quality cautions us about the possible pitfalls of an approach to urban development strongly oriented toward small business, this issue is perhaps not determinatively problematic.

Last, consider a final way that a small-business development strategy positively affects the overall prospects for effectiveness of the entrepreneurial-mercantilist strategy: Recall that, from the perspective of this strategy, small-business development is seen as a viable economic development approach in part because it diversifies the local economic base, hence increasing the dynamism and resilience of the local economy. As also noted above, this idea was derived from the theoretical work of Jacobs (1969) who contended that a diversified local economy is a key factor in perpetuating the long-term economic health of a city. Along these lines, recent evidence seems to show the soundness of Jacobs's conclusion about the importance of economic diversification.

This evidence grows out of recent econometric research conducted by Scheinkman and Schleifer. Comparing the 1956 and 1987 rates of industrial employment in 170 large metropolitan areas, they found that those areas with highly diverse industrial bases (having only 20% of jobs supplied by the top four industries) grew considerably during this period, while those with highly concentrated bases (having 60% of jobs supplied by the top four industries) declined. In addition, other aspects of their research findings highlight the importance of economic diversity for local economic growth. These economists discovered that

> those industries represented more heavily in a particular city than they were on average across the nation grew slowly. On the other hand, if the average size of companies (based on numbers of workers) was smaller than the national average, the industry grew rapidly. Finally, those industries that thrived were in cities where the other dominant industries were relatively small—apparently having a greater variety of neighbors helped. (Corcoran & Wallich, 1991, p. 103)

"The overall results," Scheinkman and Schleifer conclude, are "quite favorable" to the theoretical expectations of the Jacobs model.

Other Objectives

Because entrepreneurial mercantilism goes beyond the realm of small business development, its prospects for effectiveness also are tied to the supposed advantages of pursuing a variety of other economic development objectives.

One of these objectives is to increase the degree of *local* ownership and control of productive assets and enterprises. Central to the proposition that this objective, if achieved, will strengthen the city's economic vitality is the idea that locally owned enterprises stimulate the local economy to a greater degree through, for example, their spin-off purchasing practices and the local recirculation of their profits. Economic analyses tracking the effect of certain kinds of businesses on the local economy do provide some evidence for this claim. For example, a study of mergers in Nebraska found that 45% of companies bought out by conglomerates changed to centralized purchasing of supplies and materials (Alperovitz & Faux, 1984, pp. 146-147). Moreover, in another line of research, studies demonstrate the destructive effects of a fast-food restaurant franchise of a large nonlocal corporation on a local economy.[35] For example, a McDonald's restaurant in Washington, D.C., was found to be exporting $500,000 of its $750,000 monthly gross revenues out of the local area (Morris, 1982a, p. 6).

Picking up on this analysis, Gunn and Gunn (1991, pp. 25-37) conducted a more detailed economic analysis of a hypothetical McDonald's and concluded that, compared to a locally owned and operated independent restaurant or diner, the development of a new McDonald's in a community "can turn out to be a suction pump extracting the essential material basis for further development" from a community (p. 34).[36] These franchises "drive local competitors out of business," while the chain's corporate-linked supply practices limit local purchasing, and a good share of the profits flow to the distant corporate headquarters. The implication drawn from these studies is that the destructive consequences of nonlocal ownership can be extrapolated beyond the food service realm to other sectors of a community's economy.[37]

Another key economic development objective of entrepreneurial mercantilism is "import substitution." Recent international experiences compel us to question the wisdom of this entrepreneurial-mercantilist objective: The import substitution strategy pursued by several developing countries in the 1950s and 1960s, especially those in Latin America, is today viewed as largely a failure. In contrast, those developing nations that eschewed import substitution in favor of export-oriented development efforts—for example, South Korea and Taiwan—have experienced high rates of economic growth (see Todaro, 1985, pp. 408-409).

Nevertheless, despite the conclusiveness of this evidence, the applicability of these experiences to our subject remains dubious. Jacobs (1984), a defender of the idea that import substitution or (what she calls) "import replacement" is crucial for economic growth, explains that the tragic mistake made by development experts calling for the replacement of imports in developing countries is their "preoccupation with national economies." Because of this preoccupation, they

> do not think of import replacing as the *city process* [italics added] it is. Thinking of it instead as a national process, they often advocate that already completely developed factories . . . be set down arbitrarily any place—in little towns, in the countryside, usually wherever jobs are badly needed. All this, although it goes under the name of import-replacing or import substitution, is remote from the realities of where and how the feat of replacing imports is successfully pulled off in the real world; so remote from the realities that *such schemes can, and indeed have, bankrupted countries instead of helping them to prosper* [italics added]. (pp. 43-44)

What is decisively important for a nation's economic vitality, Jacobs argues, is the ability of its city economies to replace or substitute for imports; it matters little whether the imports being substituted for happen to be domestic or foreign (Jacobs, 1984, p. 43). Hence, whatever the value of the *urban* import substitution measures called for by the entrepreneurial-mercantilist strategy, the misdeeds of Third World development experts would seem to have little bearing on it.[38]

The next goal of entrepreneurial mercantilism involves strengthening local economic multipliers. I noted above that because other economic development goals of entrepreneurial mercantilism (such as small-business development, local ownership, and import substitution) lead to an increase in the level of interindustry dependence, they contribute to the realization of this goal. Thus, insofar as these other development goals are economically sound (see above), entrepreneurial mercantilism's focus on strengthening multipliers would enhance the overall prospects for this strategy's effectiveness.

A more specific way the entrepreneurial-mercantilist strategy strengthens local multipliers is by encouraging manufacturing. Given the profound restructuring of city economies in the post-World War II era and the resulting hemorrhaging of urban manufacturing jobs (Kasarda, 1985), the economic rationality of this strategy would appear to be quite dubious. Nevertheless, there exists some surprising evidence to the contrary. For example, undertaking industrial modernization in "brownfield" sites has been shown to be more cost effective than modernizing at "greenfield" sites for steelmaking companies (Lynd, 1981).[39] In addition, recent administrative actions taken by the Environmental Protection Agency—including, most notably, the removal of thousands of brownfields from the list of possible Superfund sites—has made these areas considerably more attractive for industrial redevelopment (Holusha, 1995). More generally, many older urban manufacturing establishments shut down *not* because of their inherent unprofitability, but because they were not profitable enough to justify continued operation and reinvestment from large conglomerates with a wide array of investment opportunities at their disposal (see Alperovitz & Faux, 1984, pp. 147-150; Bluestone & Harrison, 1982, pp. 149-160; Lustig, 1985; Lynd, 1987a).

Whatever the merits of these insights, it is clear that the potential for reviving large, traditional heavy manufacturing industries in cities is limited. This outcome is in part a function of the more widespread decline in Fordist mass production in North America and western Europe (Scott, 1988). Yet in the post-Fordist era, the common presence of research facilities (often tied to universities) and a highly skilled workforce make urban areas attractive to advanced-technology manufacturing industries. Likewise, as alluded to above, cities in the post-Fordist era can be the site for developing "flexible production networks" of small, flexible, interdependent manufacturers (Hatch, 1986).[40]

The effectiveness of the entrepreneurial-mercantilist strategy is also wedded to potentialities inherent in the goal of "localizing" employment policies—that is, attempts by the city to maximize the number of city dwellers obtaining jobs in the city economy. Even though these policies, as noted earlier, are generally able to withstand constitutional challenges, they nevertheless face an array of problems casting doubt on their ability to be economically beneficial to cities.

For example, municipal residency requirements—the first common means by which cities pursue this goal—have been difficult to enforce or have been enforced unevenly. Moreover, cities with a relatively high cost of living often need to relax their requirements because, for certain positions, such measures are "felt to be a hindrance to effective employee recruitment and retention" (McCarthy, 1990, p. 102; see also Fletcher, 1992).[41] On the other hand, this outcome is balanced by evidence suggesting that substantial fiscal payoffs exist for cities adopting residency requirements (Eisinger, 1983; Hager, 1980).[42]

The second common means by which a city "localizes" its employment policy—requiring private actors receiving city funds or development rights to employ city residents—raises a different concern. Namely, such efforts impose direct obligations on private developers. Hence, it is unclear whether cities lacking strong economic demand for their real estate can engage in these actions on a wide scale without doing greater harm to their local economies by driving away investment (cf. Peterson, 1981).

The question of whether most cities can impose obligations on developers without harming their economies has received serious attention in the urban political economy literature. Many scholars agree with Swanstrom (1986, p. 103), who argues that almost all cities possess the "economic space" to impose obligations on downtown developers (cf. Garber, 1990; Stone & Sanders, 1987). In contrast, Smith (1989, pp. 90-91) and Muzzio and Bailey (1986, p. 14) analyze the experience that cities have had with one popular type of obligation—so-called linkage fees—and are clearly less sanguine about the possibilities for poorer communities to impose these costs on new development. In what is perhaps the definitive study of the issue to date, however, Goetz (1990) produces data generally supporting the first position. Surveying almost 300 local governments, he found that, contrary to the "conventional wisdom," a variety of communities—irrespective of their degree of fiscal and economic vitality—use development policies obligating private developers "to provide a service or public benefit in exchange for development rights" (pp. 171-183) (what he calls "Type II" policies).[43] "The data indicate that localities . . . experiencing less growth are engaging in more Type II policies than more rapidly growing jurisdictions" (p. 182).

Finance

As previously noted, to provide part of the funding for entrepreneurial mercantilism, this strategy suggests tapping a number of innovative, locally generated finance schemes.

The first and by far the most important of these schemes are efforts to draw on portions of public pension funds as a source of investment capital for local economic development. The lessening of regulatory constraints governing how pension funds can be invested spurred the increased use of these geographically restricted, economically targeted investments (ETIs) in the 1980s (Eisinger, 1991, pp. 69-70).[44] Much of this activity has occurred on the state level, in part because many states link local funds directly to state systems controlled by statewide managers. Nevertheless, many big cities do control large pension funds (Morris, 1982b, p. 197) and have started to make ETIs in their geographic jurisdictions.[45] Cities have invested these funds to build affordable local housing, provide mortgages to first-time home buyers, fund loans for small businesses, and even to augment the local supply of venture capital (see Institute for Fiduciary Education, 1989, pp. 55-64).

The crucial question, of course, is whether ETIs can, as some observers claim, "spur economic activity [in local jurisdictions], while generating sufficient income to meet the pensions' current and future obligations to their members" (Ferlauto, Stumberg, & Sampson, 1992, p. 2). So far the evidence is mixed. The ETIs made in some individual states have not turned out well.[46] On the other hand, more systematic data generated from a national survey of the largest public pension funds in the country suggest these may be isolated cases. For instance, retirement systems responding to this survey reported a high level of overall satisfaction with ETI programs. They gave 64% of ETIs a "better than neutral" or "very satisfied" rating, and only 13% a "less than neutral" satisfaction rating. Moreover, on the question of financial performance, only 4% of ETIs failed to meet their financial goals (or "benchmark" level of returns), while 37% either met or exceeded their goals (Institute For Fiduciary Education, 1989, p. 4).[47]

There are three reasons to believe that these investments will be able to provide the financial stimulus for local economic development without harming the financial integrity of their pension funds. First, funds can be insulated from risk through a variety of existing risk reduction mechanisms.[48] Second, even with the use of mechanisms for risk reduction, ETIs should be able to meet easily the (traditionally low) benchmark rates of financial return expected of pension funds. Finally, public pension funds and their beneficiaries depend on and benefit from the economic vitality of their communities, something that would be buttressed by ETIs (Ferlauto et al. 1992, p. 2; Leatherwood, 1983, p. 22). Taken together, these three reasons suggest that using local public pension fund capital to stimulate local economic development holds substantial promise for the future.

Likewise, the prospects for effectiveness of other mercantilist financial innovations appear to be equally promising. For example, "linked deposit" programs—where cities place temporarily idle public funds in private banks agreeing to invest back into the local economy—have increased the supply of capital for local economic development while allowing cities to receive the same return on their deposits (Morris, 1982a, p. 46). As one study evaluating these efforts notes, "These deposits can indeed be substantial. Typically public funds make up 10% of commercial bank deposits" (Rosen, 1988, p. 121). Furthermore, a city's sale of "minibonds"—smaller-denominated municipal bonds—also appears to be a promising way for locally generated capital to be captured by the city, while at the same time giving moderate- income citizens access to an investment opportunity traditionally benefiting the wealthy.[49]

The last mercantilist financial scheme—the proposal to create a local currency system—is, not surprisingly, more problematic. The local currency system as outlined here relies on a system of accounts (similar to bank accounts) rather than the actual printing of local money. Thus, this innovation is not inherently violative of the U.S. legal system. Never-

theless, such a system would run into problems stemming from established modes of taxation, although in Australia, where these
schemes are already in operation, solutions are being developed for this
problem (Kershaw, 1990, p. 69). An even more significant obstacle is that
systems are "totally built on trust" (Kershaw, 1990, p. 69). As a result,
until now they have thrived more in small towns than in large cities,
where peo- ple are less likely to trust one another. A fear is that members
of the system could run up big deficits in their accounts and then leave
the community. Yet in a properly designed scheme, a member could
inquire into other members' account balances before transacting with
them— presumably refusing to trade further with those who have large
deficits in their accounts. Moreover, introducing a system in small
communities within big cities—neighborhoods—partially mitigates the
problems arising from the impersonality of the city and corresponding
lack of trust among urban citizens.

UNDERSTANDING CONSTRAINTS ON FEASIBILITY

The above discussion points to the conclusion that entrepreneurial
mercantilism's prospects for effectiveness as an economic development
approach are indeed encouraging. Yet to diminish the structural dual
dependencies found in cities (hence facilitating the reconstitution of
urban regimes), these alternative strategies must be not only effective
but *feasible* in the current political-economic context—that is, they must
be capable of being implemented on a scale sufficient to alter those
structural arrangements.

As noted at the end of Chapter 2, the empirical manifestation of the
three strategies has been sporadic and embryonic. For example, no city
has implemented a comprehensive development strategy based on the
principles and theoretical logic of the entrepreneurial-mercantilist
model. However, one city—St. Paul, Minnesota—in a self-conscious
and intellectually coherent manner did *attempt* to do so in the 1980s.[50]
This effort, known collectively as the "Homegrown Economy Project"
(HEP), was ultimately unsuccessful. Yet by evincing the nature of the
constraints faced by St. Paul, an analysis of this failed attempt contrib-

utes much to our understanding of whether entrepreneurial mercantilism can be feasibly implemented in contemporary central cities.[51]

The Context: The St. Paul Regime
(the Latimer Years)

George Latimer was elected mayor of St. Paul in 1976 and served until 1989, longer than any other mayor in the city's modern history. Although he first won office by a narrow margin, in subsequent elections he was easily returned to office (Lanegran et al., 1989, p. xv). Not surprisingly, given its interest in building a "homegrown" economy, the urban regime of the Latimer years clearly had elements of a "progressive" or "populist" regime form (Margolis, 1983; Shearer, 1989). Moreover, as Frieden and Sagalyn (1989) explain, Latimer himself was a

> labor lawyer who had run as a Democratic-Farmer-Labor candidate with strong ties to trade unions and almost none to the business community. He had made no promises to the downtown business establishment and had never even met its leaders. As he put it, "We didn't drink in the same bars." (p. 121)

Nevertheless, St. Paul's regime under Latimer's leadership essentially fits the profile of what has been identified analytically in this research as the "dominant urban regime form" (see Chapter 1). Both of the attributes of this "dominant" political pattern were manifest empirically in the city: (a) Key officials of the local state worked relatively closely with downtown (land-based) business leaders in the city's "governing coalition," and (b) this coalition pursued a policy agenda that strongly favored promoting urban economic growth via corporate-center/mainstream strategies, especially focusing on downtown development.

Regarding the first of these attributes, Mark Vaught, a St. Paul attorney and longtime Latimer aide, describes the evolution of the business-government coalition, despite the initial absence of the former from Latimer's electoral coalition:

> Coming in, the business community thought he [Latimer] was the radical labor lawyer with a beard. But virtually the first thing George did when he got elected was call up Phil Nason [then the president of First National

Bank of St. Paul and one of the city's most influential business leaders], and said, "I think it's time we met so that we can see that neither of us has any horns." (quoted in Robson, 1992)

By the end of Latimer's first 5 years in office, his administration had developed a strong working relationship with the city's key business organization, the chamber of commerce (Margolis, 1983).[52]

Not unlike those in other cities, this governing coalition was forged around the second attribute of the dominant urban regime form. Soon after his election, Latimer achieved a significant reorganization and consolidation of St. Paul's planning and economic development functions "so that . . . [the city] could respond more quickly and competently to development needs" (Brandl & Brooks, 1982, p. 184). With ample financial assistance coming from a variety of sources—including the quasi-public St. Paul Port Authority, the federal UDAG program, and even private foundations—Latimer's administration worked from its start with the business community to aggressively pursue investment in the downtown area (Brandl & Brooks, 1982; Judd & Ready, 1986; Robson, 1992, p. 55). During this period the city embarked on a number of large development projects (many in the form of commercial real estate, such as office towers, hotels, retail establishments), attracting, according to one estimate, more than $1 billion in downtown investment (McCormick, 1989).

The St. Paul Homegrown Economy Project

Overview. Although it didn't attempt to pursue all of the facets of the entrepreneurial-mercantilist strategy, St. Paul's HEP clearly fit the general entrepreneurial-mercantilist model for economic development. As Mayor Latimer declared in his 1982 State of the City address announcing the HEP, his administration intended to make St. Paul "the first in a new generation of self-reliant cities" (quoted in Office of Mayor [OOM], 1983, p. 17). Furthermore, the first major planning document detailing the theory and practice of the new strategy drew heavily from the ideas of Jane Jacobs, David Birch, and David Morris—all key thinkers of entrepreneurial mercantilism. This document also listed a number of economic development priorities for the city, many of which closely resembled the major tenets of entrepreneurial mercantilism (OOM, 1983, pp. 2-3), including an emphasis on the following:

- *Local ownership* to foster economic spin-off purchasing in the St. Paul economy and to increase the likelihood that business assets remain in the area
- Creating a growing job base and increasing net tax benefits to the community *in relation to the amount of economic resources consumed*
- *Diversifying the local economy* through the proliferation of many *small businesses* rather than a handful of large ones, enhancing the stability of St. Paul's business climate and encouraging innovation
- *High interindustry dependence* in St. Paul to increase the *economic multiplier effects* of local purchasing and procurement activities
- Products and services that *directly benefit the local population* or add to the wealth of area consumers by reducing or avoiding costs resulting from the diseconomies of the existing economic environment

Viewing the St. Paul HEP from a more theoretical perspective also reveals its fundamental entrepreneurial-mercantilist nature. For example, the "thesis" of HEP stated that although the city is not a closed system, economic development policy and planning should proceed "as if the city *were* a closed system" (Morris, 1986, p. 2). By treating the city as a closed system policy makers could map and rationalize the flow of resources in the local economy and "focus on those [economic development] strategies that . . . keep as many dollars in the local economy as possible" (Morris, 1985, p. 4). Working toward this goal, the HEP broadened the focus of traditional urban economic development programs by "emphasizing the *efficient* [italics added] management of all local resources," attempting to "extract the maximum value from the community's [indigenous] human, natural, and technological resources" (OOM, 1983, p. 2).

Moreover, the "self-reliant city" that St. Paul aspired to become when it adopted the HEP principles "invests in itself and generates a significant amount of its wealth *internally* [italics added]" (Morris, 1986, p. 2). Likewise, this type of city "recycles money as much as it recycles goods. Every added cycle increases the community's wealth. Businesses are evaluated not only for the services or products they offer but for *the way they affect the local economy* [italics added]" (OOM, 1983, p. 12). In this regard, of paramount concern for the HEP was the entrepreneurial-mercantilist notion that some businesses hurt the local economy's *import-export balance*. Richard Broeker, a close mayoral adviser and the HEP's key architect, said that he and the rest of the Latimer camp realized that to the extent "we had businesses in St. Paul that would essentially be importing things into the city and exporting capital out, this was a net loss to St. Paul." In conceptual terms, he added, "our goal was to reverse this situation . . . by understanding that certain kinds of businesses were good [on this score] and others were not."[53] More concretely, this conceptual focus on improving the city's import-export balance led to a strong emphasis in the HEP on "substituting local goods and services for imported products" (Morris, 1985, p. 14)—in other words, to import substitution in various forms.

Genesis. St. Paul's comprehensive effort to implement an entrepreneurial-mercantilist development strategy began in the early 1980s. During this period, and the few years before it, local officials succeeded in attracting investment to the city by using development strategies that were basically corporate-center/mainstream. Despite this record, the mayor and his closest advisers, as Latimer later put it, "struggled with the notion of getting control of [the city's] development and our economic future." In essence, the city's key policy makers came to the realization that their city remained fundamentally economically dependent, especially to the vicissitudes of mobile capital. As a result, these public officials began to see the need to chart a new course in economic development policy, one liberating them from the need to "recklessly compete with other cities for established businesses" while, in a deeper sense, instituting a "permanent and structural rearrangement" in the local economy.[54]

The belief that St. Paul could chart such a course stemmed from the city's earlier experience with energy issues. By the late 1970s the Latimer administration had confronted head on the energy crisis gripping this country, developing an integrated municipal energy policy that included a number of specific projects designed to promote increased energy efficiency. Broeker pointed out that proactively addressing the energy problem at a time when such issues were viewed as being beyond the scope of municipal action taught policy makers in St. Paul that it was possible to locally "grab the bull by the horns" on problems that "heretofore had been exempted from local action because the federal government or state government or some other galactic authority believed that only they could do it." Broeker later developed the idea for the HEP, based, as he noted, on the premise that "if we could deal with something as global as our energy dependence, why couldn't we deal with something as global as our economic [dependence]?" They could accomplish this by, he said, "becoming more proactive in understanding the pressures and dynamics involved in a local economy and using this knowledge to craft economic decisions where we had influence to shape a homegrown economy."

Programmatic Development. When the endeavor known as the HEP was formally launched in early 1983, many individual ventures meeting the general criteria and goals set forth by the new project were already being undertaken in the city. The formal project built on these established initiatives and integrated them into a more unified and comprehensive program for local self-reliance.

Notable among these earlier ventures were energy conservation initiatives. During this time, construction had begun on a district heating system using hot water to efficiently heat a network of downtown buildings (Martin, 1989). Prior to the start of the HEP, the city also created Energy Park—an industrial and office park complex dedicated to energy-related businesses and designed for energy conservation (Webber, 1989)—and the Energy Resource Center, which financed energy-saving improvements in homes (Cullen, 1989a). In addition, before HEP began, the city already pursued actively another key HEP objective: small-business (or "enterprise") development (OOM, 1983, p. 26). Through its Business Revitalization Division, the St. Paul Department of Planning and Economic Development (PED) had extensive experience in this area, providing a range of services to small businesses "including financing, loan packaging, [business] plan assistance, and site location" (Dana, 1984, p. 1).

When the project formally got off the ground, Broeker recalled that its architects "began to identify a number of [additional] economic activities and businesses that had the characteristics of the homegrown economy." The HEP's first major planning document—its manifesto of sorts—described the details and attributes of many of these activities and businesses (OOM, 1983). One economic activity prominently featured in this original document promoted the use of plant matter (especially wood and grains) as an alternative source of fuel. Moving beyond earlier established energy initiatives, which focused on conservation, the idea here was to tap into *locally generated* natural resources and formerly unused (waste) materials that could serve as *substitutes* for *imported* sources of energy such as natural gas, fuel oil, electricity, and gasoline.

Another established economic activity embraced by the HEP was aiding the development of new (especially small) enterprises in the city. The HEP called for a number of specific strategies—including, notably, the continued provision of technical assistance and debt financing for small businesses (OOM, 1983, p. 15). Yet the HEP's key goal in this area was to expand the range and scope of the city's prior policies so that small-business development efforts would begin to have a more substantial effect on the local economy.

Reaching this end involved several innovative means. First, the creation of a seed capital fund was suggested as a means of making equity capital available "to ventures that are economically viable and can contribute to the St. Paul economy," but that are nevertheless locked out of venture capital markets (Morris, 1985, p. 12). Second, soon after the HEP officially got under way, its enterprise development efforts were given a large boost by the opening of the city's publicly sponsored business incubator—later referred to as the cornerstone of the overall HEP (Cosgriff, 1989; Dana, 1984, p. 1; Hendricks, 1989). Third, the HEP sought to promote local small-business development through the identification of new markets for these enterprises.

One interesting proposal to identify these markets was premised on the idea that large companies in the area purchased "hundreds of millions of dollars of equipment, parts and supplies annually" that represented "a potential market for many local small businesses." The HEP developed an innovative plan "to assist [these] large corporations in identifying qualified [local] small businesses to whom supply contracts can be offered which normally go to firms outside St. Paul" (OOM, 1983, p. 15). This also would be a source of import substitution for the local economy, greatly enhancing the local economic spin-off or multiplier effect stemming from the economic activities of the area's larger companies. Similarly, the HEP understood that the public sector —the federal government and local governments—could be tapped as another potential source of new markets for local small enterprises (OOM, 1983, p. 15). Finally, the HEP saw the state of Minnesota's new World Trade Center under construction in the city as the link between local homegrown businesses and the global marketplace. To strengthen this connection, the project envisioned that one or two floors of this building would be used as incubator space "for small local companies" (Morris, 1986, p. 5).

The HEP also identified a number of other homegrown economic development activities. To build on St. Paul's established energy conservation initiatives, the HEP suggested the use of underground or mined space to create an energy-efficient "city-under-the-city." To encourage technological innovation in the local economy, the HEP sought to develop contacts "not only with entrepreneurs but with inventors," (Morris, 1986, p. 3). To support the local consumption of city-made products, the project recommended a "buy local" program, with locally produced items labeled to let consumers know "that where they spent their money affected the community" (Morris, 1986, p. 10).

Finally, the HEP targeted specific businesses fitting its conception of the homegrown economy particularly well. It sought to assist the expansion plans of a wood-chipping company in the city by helping it sell its chips to a large local corporation operating in the area. The corporation, 3M, intended to use the wood chips as fuel, which would accomplish other HEP objectives by "reducing the area's dependence on imported fuels [i.e., energy] and utilizing a [formerly unused] local resource" (Dana, 1984, p. 2). A similar effort involved assisting a company that would recycle used tires for use by the city's asphalt plant as an asphalt supplement. This would eliminate the need for costly disposal measures, and "the money that was thrown into landfills [would be] now reclaimed for the community to create jobs and produce a useful product" (OOM, 1983, p. 20). An additional array of homegrown enterprises also targeted by the HEP for assistance focused on substituting for food imports by augmenting local production capacity (Dana, 1984, p. 3). Included here were a microbrewery that could use locally grown hops, a farmer's market to sell local produce, and a large energy-efficient "growhouse" for raising fresh vegetables in the city year-round (OOM, 1983, pp. 20-21).

Analysis of Impact

Despite the HEP's promising start, it produced only marginal results. This outcome, for the most part, can be attributed not to the ineffectiveness of the HEP's various economic development initiatives, but rather to the fact that, as Broeker later put it, the overall effort "never came together." Much of the planned activity outlined above was never put into practice.[55] This entrepreneurial-mercantilist effort, in short, proved to be an unfeasible urban development option and, as such, did not have a significant effect on the structural context of the urban regime.

At one level, the reasons *why* this effort never came together can be characterized as idiosyncratic and hence case specific: Less than 18 months after it began, HEP lost its chief architect, Deputy Mayor Richard Broeker—Latimer's key aide and arguably the second most powerful political leader in the city—when he left the mayor's office, which housed the HEP under his direct supervision. This "badly damaged" the "clout and legitimacy of the Project" (Morris, 1986, p. 11). Moreover, with Broeker and another key HEP staffer gone, the location of HEP was moved from the center of power in city government, the mayor's office, further weakening the status of the HEP. Then the HEP received its final, fatal blow: Mayor Latimer decided to run for governor of Minnesota the next year.[56] This effort resulted in an enormous drain on the attention and energy required for the implementation of the HEP—a new, comprehensive reorientation in the city's development efforts.

Although small pieces of the HEP remained intact well into the 1990s, the formal, comprehensive effort halted in early 1986. Latimer stepped down as mayor in 1989. His replacement, Jim Scheibel, a neighborhood-oriented city council member, wanted very much to avoid the shadow of the popular, charismatic Latimer. Hence, he expressed little interest in reviving the HEP, which was closely associated with Latimer, even though he may have been broadly sympathetic to the idea.

Thus, from a superficial viewpoint, the reason why the effort to build a homegrown economy in St. Paul never came together can be attributed largely to idiosyncratic circumstances such as key personnel changes, unfortunate timing, and the lingering effects of an extraneous political rivalry. Clearly these factors do explain much of the outcome. Yet by undertaking a deeper analysis of the St. Paul experience, we can gain a better understanding of the more general conditions likely to limit the feasibility of an entrepreneurial-mercantilist strategy across a variety of urban contexts.

Empirical investigation identified three broad sources of impediments inhibiting the development of the HEP—political, institutional, economic.

Political Factors

The subtext running through the entire story of the HEP's demise sketched above was that the project never gained strong, continuing political support from the city's leaders after Deputy Mayor Broeker's departure from local government. Given the city's strong mayor form of government, the most important of these leaders was Latimer himself. As David Morris, in retrospect, noted: "The effort (HEP) was viewed by many as . . . [having] little support from the mayor" (Morris, 1986, p. 11).

What is striking, however, is that despite Latimer's reluctance to support the HEP with enthusiasm and his concomitant unwillingness to devote his considerable leadership talents to it, evidence suggests that he believed—rightly or wrongly—in the basic economic rationality of the approach. Latimer, after all, was convinced by his aides to initiate the unconventional HEP and make it a major focus of his administration, at least for a while. "I remember," he recalled, "that when I read Jane Jacobs for the first time and she talked about import substitution and all that, I realized that's what [urban economic development] was all about." (personal communication, winter 1991-1992)

"The strategy is generally sound," he continued. "It is clear that cities [pursuing it] can almost always gain a positive economic return on their investments. If you [do] fail, the nice thing about [this type of strategy] is that you haven't left craters in the earth, canyons of empty spaces, and inefficient investments—you have not really screwed everybody up. So there is no downside in the ugly way that over-investment and mis-investment can do in other approaches."

Nevertheless, as a large body of research demonstrates, economic rationality rarely shapes the urban governmental agenda; political rationality does (see Stone, 1987; Swanstrom, 1988). Latimer, an astute and perceptive politician, saw myriad political barriers inhibiting the development of the HEP in St. Paul. This type of strategy, he came to realize, "requires massive intervention" by a city government, but it is "not sexy politically." He also saw that it is a long-term strategy requiring, he said, "10 to 20 years of investment to see the full results," but "mayors and council people don't get elected in 10 or 20 years." Likewise, as the local economy suffered the "loss of thousands of industrial jobs," he felt political pressure to respond in a quicker and more substantial and visible way than possible under the nascent HEP.

Latimer, it should be noted, was not alone in understanding the political difficulties burdening this alternative strategy. For example, one key middle-level economic development official in the city charged with the task of implementing HEP activities commented about its inability to garner the press attention so important for politicians. He noted that, in urban economic development, you get "front-page press attention for a large downtown project," yet when you help a multitude of small businesses "you don't get that same press attention." Broeker made a similar point: The HEP involved "a million little things, not one big thing like a new high-rise building that does nothing for your local economy but attracts a lot of media attention." Yet "the newspapers and the rest of the world do not like a million little things," he continued, "they want the big bang, the headline."

Mayor Latimer also seemed to understand that, as pointed out in Chapter 2, the general weakness of the local state makes it politically necessary for public officials wishing to accomplish complex urban policy tasks (such as the building of a homegrown economy) to be able to draw on extrastate resources lodged in the local community (cf. Stone, 1989). This had been his administration's formula for success several times before (see Margolis, 1983, p. 46; Osborne & Gaebler, 1992, pp. 25-26). In the St. Paul context, the area banks and corporations and, to a lesser extent, labor unions and area foundations held the necessary resources. Although the HEP was not in direct conflict with the political interests of any of these institutional actors, Latimer nonetheless believed that its unconventionality would make the task of mobilizing these actors in support of the HEP very difficult. "The idea," he reflected, "is hard to sell to the big institutions that have traditionally been at the heart of a city like St. Paul." As a result of this belief, the mayor decided not to try to sell it to them.

Thus, in essence, Latimer's political instincts led him away from the pursuit of any major push for a reorientation of his city's development efforts toward the entrepreneurial-mercantilist model. The strong corporate presence in the city's governing coalition reinforced his inclinations in this direction. Further reinforcement came from the fact that, during this period (the early to mid-1980s), St. Paul's more traditional economic development efforts (which emphasized the physical redevelopment of the downtown area) appeared to be working well (see

Judd & Ready, 1986, p. 242).[57] Without strong mayoral support, however, one key actor later observed that the HEP "never got the kind of political cachet" it needed to thrive.

Institutional Factors

It is plausible to conjecture that a profound impediment would arise because cities lack the necessary institutional capacity—organizational strength, managerial competence, and policy expertise—to implement such a sophisticated urban development strategy. This was less of a factor in the St. Paul case, however, because the effort remained limited. The problem was not capacity, but recalcitrance: The HEP faced significant opposition from the city's bureaucratic apparatus. This opposition to the basic idea was especially pronounced in the more traditional departments in city government such as Public Works, and Parks and Recreation. To a lesser, but ultimately more consequential degree, this posture was also exhibited by the agency principally responsible for implementing the HEP: the city's Department of Planning and Economic Development. One important participant said that the leaders of this department "were actively opposed to the concept" (Morris, 1986, p. 1). James Bellus, director of the department at the time, admits that he "was not a great fan of the HEP" and that he was "a naysayer at the time" and "always has been." (personal communication, winter 1991-1992)

A deeper look at this issue, however, reveals that although key officials at St. Paul's Department of Planning and Economic Development had serious misgivings about the HEP itself, they—like Mayor Latimer—were broadly sympathetic to the essence of the approach. One of these key officials was Alan Emory, a longtime employee of the department who had risen through the ranks to become a deputy director. He found the approach to be sound because it emphasized stimulating indigenous development over the attraction of businesses from other places. The latter, he noted, was a waste of time and resources and ultimately counterproductive. More generally, he argued

If a city is going to do economic development, it should do it this way. You should try to get the maximum economic benefit out of [development] by paying attention to things like multiplier effects and trying to understand how businesses affect the local economy.

Even Bellus, the well-respected chief of the department, expressed similar sentiments:

> I think [its approach] was exactly right. I really do believe that the idea of a homegrown economy and independence and self-sufficiency are the absolute right philosophies on which to build an economic development program for a community. Absolutely—I have no doubt about that.

Why, then, did the HEP face substantial bureaucratic opposition? To answer this question, one must understand that the effort pursued in St. Paul was simply one possible version of the general entrepreneurial-mercantilist model. The city's particular version put the goal of "efficient resource utilization"—especially via the use of alternative technologies to tap underused local resources for the local production of commodities such as energy and food—in the forefront of its economic development program. In a word, then, the HEP privileged the "green" elements of the entrepreneurial-mercantilist model over the more straightforward aspects of this economic development approach.

This prioritization limited the appeal and the legitimacy of the HEP among most of the city's development experts, who were uncomfortable with these very nontraditional ideas. Even David Morris, a specialist on alternative energy paths who was mainly responsible for the "greening" of the St. Paul effort, acknowledged this fact: "There was a sense that the concept [of the HEP] was hippie-ish," and many city officials "felt that it was really at the margins and that you are not going to be able to have a city policy around this." Similarly, Bellus added that "to the average city employee, the HEP seemed flaky," and many of its individual endeavors, such as hydroponic gardening "seemed far-fetched." Broeker, likewise, recalled "some giggling" on the part of some city officials during planning sessions for the HEP.

Economic Factors

A third source of circumstances inhibiting the implementation of the HEP were of an economic nature. The first of these also can be traced to the makeup of the particular example of entrepreneurial-mercantilism pursued in St. Paul. As alluded to above, efforts designed to promote local self-reliance in energy through efficient resource utilization were at the center of the St. Paul version of this strategy. This approach would have made enormous economic sense had energy prices continued to soar throughout the 1980s. But these prices stabilized in the early 1980s and, by the mid-1980s, fell precipitously (for example, the price of oil dropped from more than $28 a barrel in November 1985 to less than $10 in April 1986). This change in the economic environment caused a major thrust of St. Paul's strategy to appear unnecessary, inhibiting the development of overall HEP effort.

More fundamentally, the general economic dependency of the city also inhibited the growth of the HEP—although, it should be noted, the HEP never expanded to the point where the extent of this factor could be understood fully. In St. Paul, as in most older industrial cities, two related trends defined this dependency: economic restructuring and suburbanization. From the late 1970s to the late 1980s a number of the manufacturers in the city closed or relocated plants, curtailed production, or expanded operations elsewhere.[58] As for suburbanization, the city faced not only that but also a situation where almost all of the economic and population growth in the region was occurring in the suburbs west of Minneapolis. This created a strong economic and demographic pull away from the city of St. Paul, which lies to the east (Robson, 1992, p. 54).[59] As several key observers noted, these trends made it difficult for many HEP ventures to get off the ground.

The economic dependence of the city acutely inhibited the development efforts of the HEP. This acuteness arose as this general dependency interacted with a second economic reality: the antilocal biases of the larger economic system. As Broeker pointed out, the principal policy actors involved in the HEP came to realize how much "the economic system is rigged toward bigness and globalized solutions and against the *localism* which was at the heart of the HEP." As a particularly stark example of the phenomenon, he cited the distribution difficulties faced by a beer made in a local microbrewery: "This beer . . . is wonderful, but no [retailers] wanted to carry it because they were on the well-oiled Miller and Budweiser distribution pipelines." In short, Broeker continued, these retailers found it easier and more convenient to continue their established and more predictable business relationships "than to buy homegrown." This same general bias, moreover, plagued a variety of the HEP's other economic development initiatives, making it more difficult for the city to implement these perhaps ultimately effective development measures.

Summary

The St. Paul experience with the HEP points to some of the key impediments limiting the feasibility of a city's efforts to reorient its economic development strategy in an entrepreneurial-mercantilist direction. Particularly salient are political barriers. As this experience showed, the reorientation of a city's economic development efforts in this manner requires strong political leadership to guide the process. Yet serious political disincentives are at work, inhibiting leaders from taking the necessary action. Institutional factors, on the other hand, were less of an impediment in the St. Paul case, largely because the effort remained modest. The St. Paul case did, however, point to the strong role that economic factors play in limiting a city's ability to implement such a strategy. In part, these economic factors stemmed from the unique type of self-reliance strategy pursued in St. Paul, with its heavy emphasis on measures to promote efficient energy utilization. More generally, however, the antilocal biases of the larger economic system made the condition of economic dependence experienced by the city an especially acute impediment.

Despite the evidence indicating the possible effectiveness of this strategy, the above analysis shows how that effectiveness can be compromised by difficulties cities face in attempting to implement it (on a scope large enough to alter the structural context of urban regimes). The myriad impediments, taken together, are indeed formidable. In light of this finding, our final chapter will explore the possibilities for and likelihood of overcoming these impediments.

NOTES

1. Sections of this chapter appear elsewhere in a slightly revised form (see Imbroscio, 1995a).

2. Jacobs's theory here complements the earlier work of Joseph Schumpeter (1942/1976), which emphasized the key role in the economic development process played by innovation or "new combinations" brought about by entrepreneurs (Malizia, 1985, pp. 39-41).

3. To illustrate this, Jacobs (1969, p. 86) contrasts Manchester and Birmingham, England, in the mid-19th century. The efficient Manchester, with the "stunning efficiency of its immense textile mills," proved unable to adapt to changing economic circumstances and subsequently declined. Birmingham, with its incomprehensible array of small enter-

prises not rationally and efficiently consolidated, retained considerable innovation and "remained economically vigorous and prosperous" (Jacobs, 1969, p. 89).

4. This perspective is also broadly consistent with a Gandhian approach to economic development (see Gangrade, 1989).

5. This third advantage arises because they are often "labor intensive" rather than "capital intensive" operations. As a result, these businesses offer numerous job opportunities that require only limited education and skills, and as such fit the employment needs of the generally low-skilled urban population (Long, 1987).

6. Numerous cities have offered technical assistance to small enterprises, including Oakland, California, and San Antonio, Texas (Schweke, 1983; see also Bowman, 1987a). Many other cities also have financial assistance programs for small companies (see Bowman, 1987a). New York City, for example, maintains a small-business loan fund of $10 million (see Lueck, 1992), while at the other end of the urban spectrum, the small city of Grand Forks, North Dakota, has a financial assistance program that provided $3.9 million to businesses over the past 4 years (Terry, 1992).

7. This activity is already quite commonplace at the state level. As Eisinger reports (1991, p. 67), 23 states have started a total of 30 state venture capital programs. At the municipal level, Bowman's (1987a, p. 4) survey of 322 cities found that 25 (7.8%) had used venture capital as an economic development tool in the past, and 85 (26.4%) said they are likely to use it in the future.

8. "Micro loans" to encourage people to become entrepreneurs were a component of the development plan of the administration of Harold Washington, "Chicago Works Together" (see Wiewel & Rieser, 1989). Some states, including Washington, allow the unemployed to use their benefits (received in lump-sum payments) as start-up capital for microbusinesses (McCarroll, 1992), and others, including Maryland, provide micro business loans to welfare recipients (Clark, 1992). Entrepreneurial strategies like these also are frequently employed in Western Europe (see Friedman, 1986, p. 43; Meehan, 1987, pp. 137-139).

9. Publicly sponsored small-business incubators have been established in the cities of St. Paul, Minnesota (see case study below) and Chicago (see Giloth, 1991, pp. 112-113; see also Weinberg et al., 1991).

10. City officials in Eugene, Oregon, in conjunction with a private bank and the Lane County Private Industry Council, established a "Buy Oregon" program: "The program starts by contacting businesses to determine which products they are buying from nonlocal sources. Detailed specifications are then circulated to a list of potential local suppliers." Finally, when "a buyer has indicated an interest in a particular supplier," purchasers and sellers are brought together through the program (Persky et al., 1993, p. 21).

11. "Chicago Works Together" set a goal to increase the city's local purchasing by $80 million (Judd & Ready, 1986, p. 232). Chicago favors local suppliers by considering their bids competitive even if they are 2% higher than nonlocal firms (Persky et al., 1993, p. 21). Other cities, such as Detroit and Livermore, California, have considered the bids of local firms to be competitive even if they are up to 5% higher than nonlocal firms (Morris, 1982a, p. 46).

12. This approach was an important part of the "Homegrown Economy Project" undertaken in St. Paul, Minnesota. See case study below.

13. Numerous cities have employed resource conservation strategies like these. One of the most innovative and aggressive is Davis, California (see Shavelson, 1990). Also see the case study of St. Paul, Minnesota below. Energy conservation is often emphasized

here because, as Morris (1982b, pp. 188-189) notes, "The dollar spent for energy in a locality disappears from it rapidly." Studies show that on average only 10 to 15 cents of each dollar spent on energy remains in the city economy.

14. Manufacturing industries generally have a higher multiplier effect on the local economy because they are "export" industries. Yet analysis indicates that many export industries in nonmanufacturing sectors may not increase local multipliers in cities because they fail to stimulate the rest of the local economy (Long, 1987, pp. 196-197; Stanback & Noyelle, 1982, pp. 140-142; see also Patton & Markusen, 1991).

15. Several cities, including New York, Baltimore, and—most aggressively—Chicago (with its Planned Manufacturing Districts) have used zoning controls to protect manufacturing areas from displacement by encroaching commercial and residential land uses (Giloth & Mier, 1989, p. 192). Chicago also set up industry task forces to develop ideas to preserve its steel and apparel manufacturers (Giloth & Mier, 1989, p. 196-202; see Markusen, 1988).

16. St. Paul, Minnesota, again offers an interesting example, as this type of evaluation was an important conceptual basis of its "Homegrown Economy Project" (see case study). More generally, the attempt to increase interindustry dependencies—"the degree of 'linkagedness' of a firm to the rest of the local economy"—was the "essence" of the plan developed by Luria and Russell (1982, pp. 170-171) to "rationally reindustrialize" Detroit in the early 1980s.

17. Several states have begun tapping this capital pool to stimulate their economies. These geographically restricted, economically targeted investments (ETIs) have been employed in Alabama, which invested pension-fund money in an aircraft parts plant and a pulp mill; Connecticut, a maker of firearms; Kansas, a steel mill; Colorado, car washes and biotechnology companies. Locally, New York City invests a portion of its pension funds in a small-business loan program as well as other local development efforts (Stevenson, 1992). ETIs also have been employed in Philadelphia, San Jose, Houston, Milwaukee, Chicago, San Francisco, Baltimore, and Hartford. A 1989 survey of the nation's 126 largest public pension funds found that 41 of the 99 responding make ETIs totaling nearly $7.3 billion in investments (Institute for Fiduciary Education, 1989).

18. A number of cities, including Santa Monica, the District of Columbia, Pittsburgh, and Chicago "link" their bank deposits (see Morris, 1982a; Rosen, 1988; Schweke, 1983).

19. A few dozen local governments and authorities have issued "minibonds" over the years, in places such as New Orleans; Philadelphia; East Brunswick, New Jersey; Ocean County, New Jersey; and Rochester, New York. New York City recently started a "minibond" program in which it sold more than $100 million worth of "zero-coupon" minibonds (Fuerbringer, 1992; Morris, 1982b, p. 196).

20. Local currency systems are found in many Canadian and Australian cities, including Brisbane, Sydney, Melbourne, Calgary, and Vancouver (Landsman Community Service, 1989). In the United States, local currency systems exist in Great Barrington, Massachusetts, and Ithaca, New York, and a small demonstration project can be found in Los Angeles (McGuire, 1992; Nakano & Williamson, 1993). For a different proposal for a local currency based on local transactions for energy resources, see Benello (1988).

21. Many cities have adopted residency requirements for public employees (see Eisinger, 1983; Greene & Moulton, 1986), including Boston and Washington, D.C. (Fletcher, 1992). Likewise, measures requiring local employers to hire (or consider hiring) city residents are also common. Obligations are usually incurred when companies receive public funds for development efforts, of which the Chicago "first source" program discussed by Stanback and Mier (1987, p. 17) is an example. Dreier (1989, p. 51) reports

that Boston extended its requirements that developers fill half of their construction jobs with city residents beyond publicly funded projects, imposing these same requirements on "the much larger number of private projects."

22. Although these concepts are usually referred to in the context of nations trading in the international system, they apply equally well to geographical sub-units of national economies (e.g., states, regions, and cities) trading with other sub-units. For a general discussion of "comparative advantage" and "gains from trade," see Alt and Chrystal (1983, pp. 98-100).

23. This by no means implies that all trading and economic contact with the outside world is thoroughly discouraged by this economic development strategy; entrepreneurial mercantilism should not be characterized as a strategy of economic "autarky" or complete self-sufficiency (cf. Meehan, 1987, p. 135). Yet it *is* clear that, in its enthusiasm to create development from within the local economy itself (including its bias toward using locally generated capital and other locally generated economic resources to promote it), the philosophical thrust behind this approach leaves it reluctant to make an obeisance to the orthodox postulate that a more open economy is always more robust.

24. See Supreme Court cases *Hughes v. Alexandria Scrap Corporation* (1976), *Reeves, Inc. v. Stake* (1980), and *White v. Massachusetts Council of Construction Employers, Inc.* (1983). Also see Clark (1985, pp. 103-110) and Mandelker et al. (1990, p. 396).

25. The Privileges and Immunities Clause states, "The Citizens of each State shall be entitled to all Privileges and Immunities of Citizens in the Several States." The Supreme Court, in a case challenging a municipal ordinance in Camden, New Jersey, requiring contractors working on city projects to hire city residents for 40% of their jobs, noted that it "imposes a direct restraint on state action in the interests of interstate harmony. This concern with comity cuts across the market regulator-market participant distinction that is crucial to the Commerce Clause." See *United Building & Construction Trades Council of Camden County v. Camden* (1984), reprinted in Frug (1988, p. 471).

26. See the Supreme Court's ruling in *United Building & Construction Trades Council of Camden County v. Camden* (1984).

27. See 6th Circuit case *Wardwell v. Board of Education* (1976) and the Supreme Court case *McCarthy v. Philadelphia Civil Service Commission* (1976).

28. An "establishment" is simply a place of business, which can be an independent operation or a branch operation of a much larger company. Many smaller establishments are actually in the latter category, yet in Birch's research design these operations are defined as "small businesses." As a result, the overall role played by small business in job generation is exaggerated, as many of these small establishments are actually parts of much larger businesses (Eisinger, 1988, p. 236). A "firm," on the other hand, is a separate business that may or may not operate at multiple "establishments."

29. Birch's more recent work (1987) also is more ambivalent about the role of small business in job generation. Although he concludes that "very small firms [those with fewer than 20 employees] have created about 88% of all net new jobs in the [1981 to 1985] period" (Birch, 1987, p. 16), he acknowledges that this fluctuates considerably according to the period under examination. In particular, during periods of recession, larger companies tend to cut back their work forces substantially, so the number of net jobs created by smaller businesses swells.

30. A later study by the Small Business Administration (SBA) came to a similar conclusion (cited in Fisher, 1988, p. 168). Looking at the years 1976 to 1982, it found that independent small companies with fewer than 100 employees accounted for 52.6% of net new employment. The SBA study also showed that very small companies (those with

fewer than 20 employees) were even more potent job generators. Although accounting for only 20.5% of total employment in 1976, from 1976 to 1982 they generated jobs at nearly double that rate (38.5%).

31. Although this data file is very large, it has numerous deficiencies (Vaughan & Pollard, 1986; White & Osterman, 1991). For example, the file includes only those companies that seek credit, and so omits 20% of all business establishments in the nation. Vaughan & Pollard (pp. 124-125) also note that the file is not automatically updated, that it has steadily covered an increasing number of establishments since the late 1960s, and that its treatment of mergers, acquisitions, and franchises makes it difficult to track individual enterprises.

32. A California study, for example, found that small businesses supplied about 80% of net new jobs from 1976 to 1979, while a Wisconsin study reported that small businesses generated almost 70% of new jobs from 1977 to 1978 (Eisinger, 1988, p. 238; Zipp, 1991, p. 10).

33. Data show that for about 30 years after World War II, the average size of U.S. companies grew. In the 1970s or early 1980s, however, the average size of companies began to decrease. This trend holds for both the service sector and the manufacturing sector (see Case, 1989).

34. Drawing on the experiences of flexible networks of small enterprises operating in north-central Italy, Hatch (1986, pp. 16-18) outlines the bountiful economic development possibilities for New York City engendered by externalization and the accompanying economic agglomeration. (On the Italian experience, see also Piore & Sabel, 1984; Scott, 1988.)

35. These fast-food franchises, such as McDonald's and Burger King, though perhaps owned locally, are more accurately understood as branch operations of large, nonlocal corporations.

36. As the basis for this claim, the Gunns show that 77% of the "social surplus" from a McDonald's restaurant leaves the local area (Gunn & Gunn, 1991, p. 34). Social surplus, they explain, "is the difference between the net product of a society and the consumption (in individual and collective forms) that is essential to maintain those who do the producing" (p. 3).

37. One important area is retail merchandising. See, for example, Gunn and Gunn (1991, pp. 126-127) on the economic effects of a Wal-Mart store on a local economy.

38. In contrast, in their discussion of more city-oriented attempts to promote import substitution in the United States, Persky et al. (1993) point out that many advantages and benefits can stem from these still nascent local programs, currently operating in such places as Chicago and Eugene, Oregon.

39. "Brownfields" are often urbanized, older industrial areas such as Pittsburgh or Youngstown; "greenfields" are areas that have hitherto been rural. Citing two comprehensive federal studies, Lynd (1981, pp. 28-29) shows that "it is cheaper for the individual steel company to modernize in existing locations than in new, greenfield sites." And though these studies were done in the late 1970s and early 1980s, Lynd notes, "There is no reason to suppose that the comparative figures for the cost of brownfield and greenfield modernization will change significantly in the foreseeable future" (p. 30).

40. For example, Hatch (1986) forwards a proposal to reindustrialize New York City through the promotion of these flexible production networks, and Fosler and Ehsani (1992) demonstrate in a study how these networks could boost the fortunes of Baltimore's machining industry.

41. One city in this category was Washington, D.C. It recently dropped its residency requirement for police and other public safety employees for this reason (Fletcher, 1992).

42. Eisinger (1983) estimated, for example, that New York City in the mid-1970s would have gained approximately $56 million in tax revenues alone if it had reinstituted its residency requirement. In addition, he adds, "The [fiscal] implications of a residency law are for some [other] cities even more weighty," as these other cities tend to have a "proportion of suburb-dwelling civil servants [that] is often greater than in New York" (p. 95).

43. "Type II" development policies include, for example, programs requiring that local citizens be offered employment (our concern here).

44. According to recent estimates, state and local public pension funds control more than $875 billion in assets (Conn, 1992). A survey of the 126 largest public pension funds in the nation found that 41 of the 99 responding to the survey made ETIs totaling nearly $7.3 billion in investments.

45. The survey by the Institute for Fiduciary Education (1989, p. 2) found that 51.7% of statewide systems reported making ETIs, while 26.8% of local (county or municipal) systems made this kind of investment.

46. Alaskan fund managers, for example, reported to the Institute for Fiduciary Education (1989, pp. 7-8) that heavy investment of pension fund assets for in-state commercial and residential mortgages was a "disaster" because of the subsequent oil price crash, "leaving the funds with a legacy of 40% nonperforming loans (approximately 30% delinquent, 10% foreclosed)." Likewise, in Kansas a public pension fund lost about $200 million from ETIs when it invested in such things as a steel mill that closed and a savings and loan association that failed (Conn, 1992; Stevenson, 1992).

47. The rest (59%) either did not know if their financial goals had been met (largely because their ETI programs have been in place for too short a time period to evaluate their financial performance) or did not answer the question (Institute for Fiduciary Education, 1989, p. 4). Specific data from one city (New York) using its pension funds for economic development purposes confirm these more general survey results: The average investment yield of the ETI program of its City Employee Retirement System over the past 5 years was 13%, compared with only 12.1% for the overall portfolio (Stevenson, 1992).

48. Fund managers using ETIs can (a) purchase debt instruments with federal guarantees or package their loans for resale to other investors; (b) channel ETI money through an intermediary such as a local bank that analyzes investment opportunities, monitors performance, and often pools risk over multiple investors; or (c) participate in public or private insurance programs, join private or state-sponsored loan guarantee programs, or form limited partnership arrangements with private sector firms (see Ferlauto et al., 1992; Institute for Fiduciary Education, 1989).

49. New York City's recent minibond sale illustrates this point: Targeted through advertising to city workers and retirees, it was deemed to be immensely successful, as the city was able to sell double the amount in bonds it planned to offer (selling $106.6 million).

50. In a less comprehensive way, various elements of the economic development plan, "Chicago Works Together," pursued during the mayoral administration of Harold Washington (1983-1987) also could be viewed as consistent with the entrepreneurial mercantilist strategy (see Giloth, 1991; Judd & Ready, 1986).

51. Except where otherwise noted, the analysis provided here draws on a series of interviews conducted with key informants both by telephone and in person during the winter of 1991-1992.

52. One manifestation of this strong political relationship between the Latimer administration and the local business community was the declining fortunes of Republican mayoral candidates in St. Paul. Although the city elected four Republican mayors in the 16 years before Latimer's initial victory, Republican mayoral candidates won just 13% and 27% of the vote in the mayoral elections of the early 1980s (Margolis, 1983, p. 46).

53. Clarifying, Broeker cited an example:

> When you've got a new McDonald's franchise opening in town and they paid minimum wage and a good chunk of the money went to wherever the headquarters was located, and they are not using any local food or any local produce, then it isn't much of an advantage to have a McDonald's in the community from an economic standpoint. Little investment for the local economy is generated from this enterprise, and it is probably wiping out a couple of family restaurants that . . . took better care of the people they employed and who bought locally and who kept the money in the community. . . . Other things being equal, from an economic standpoint you want the latter type of business and not the former.

54. The first of these quotes is taken from Latimer's 1982 State of the City address, as quoted in Judd and Ready (1986, p. 238); the second from an interview with Richard Broeker.

55. For example, no large-scale effort to tap local plant matter as alternative fuel was undertaken; a seed capital fund was developed, but it was small and was started long after the comprehensive effort ended; the innovative plan to link large area companies with smaller local businesses was not followed through on; portions of the city's World Trade Center were not used as incubator space as intended; the buy local program was not sustained; major plans for mined space were dropped after opposition arose; the tire recycling operation was opened, but in the northern section of the state; and the grow-house experiment was not tested in an urban setting (Morris, 1986).

56. Latimer lost badly in the Democratic-Farmer-Labor (DFL) primary to the incumbent governor, Rudy Perpich, by a margin of 57% to 41%.

57. Over the longer term, the picture subsequently changed: Many of the city's development projects undertaken in the 1980s turned out to be financial disasters, costing the public millions (see Carr & Wieffering, 1986; Fiedler, 1990; Frieden & Sagalyn, 1989, pp. 219-220; Papatola, 1990; Robson, 1992, p. 55). Jolted by the weakened real estate market, by March 1991 the city's economic development financing arm, the quasi-public St. Paul Port Authority, experienced at least 28 outstanding portfolio defaults. As a result, the Port Authority's bond rating was downgraded three times by Standard & Poor's within less than a year, eventually claiming a standing just above junk bond level (Oxnevad, 1991; Robson, 1992, p. 55; Saint Paul Port Authority, 1991).

58. For example, according to a PED survey, from 1984 to 1989, St. Paul's manufacturing base declined by 30%, a loss of 2,054 jobs and 39 companies (Fiedler, 1990, p. 34). Moreover, many of the manufacturing companies that closed, relocated, curtailed production, or expanded elsewhere during the Latimer years were some of St. Paul's most venerable businesses. These included American Hoist and Derrick, Whirlpool, Brown and Bigelow, Burlington Northern, and 3M.

59. From 1980 to 1987, 53.7% of the region's 159,162 new jobs were in the suburban portion of Hennepin County (Fiedler, 1990), which is largely to the west and southwest of Minneapolis. All but one of the region's fastest-growing jurisdictions during this period were also suburbs lying to the west of Minneapolis: Eden Prairie (which grew 115%), Maple Grove (which grew 75%), and Plymouth (which grew 52%). The only exception was Eagan (which grew 105%), a suburb to the south of St. Paul (Walden, 1989).

THE COMMUNITY-BASED STRATEGY

This chapter examines "community-based economic development"—a second alternative urban economic development strategy. Community-based economic development, like entrepreneurial mercantilism, has the potential to bring about the diminution of the structural dual dependencies in central cities (hence promoting the reconstitution of urban regimes in ways lessening political inequality).

As we shall see, much of the entrepreneurial-mercantilist strategy is not inconsistent with this second strategy. For example, the task of building in cities an indigenous capacity to achieve economic vitality is also of central concern here, but a community-based approach suggests a fundamentally different direction toward realizing this goal. Most important, unlike entrepreneurial mercantilism with its generally individualistic (or "atomistic") comportment,[1] the community-based strategy has at its intellectual and theoretical heart the notion that urban economic development is to be a collective affair (see Wiewel, Teitz, & Giloth, 1993, pp. 96-97), involving efforts undertaken *by and for* some (essentially private) "collectivity" known as the "community." As Rita Mae Kelly (1977) points out, "The community economic development movement stresses cooperation and initiative by the private sector, but instead of stressing individual effort only or primarily, it stresses *the role* [italics added] of the total community" (p. 137). In this collective approach, adds Gar Alperovitz, what "becomes important" is the idea that "wealth is held for *the benefit* [italics added] of the broader community" (Alperovitz, 1991, p. 1; see also Alperovitz, 1990).

The discussion below follows the format used in the previous chapter. I first sketch the vision projected by the community-based strategy and illustrate its substantive core. To better evaluate this strategy's potential to be the vehicle for lessening the dual dependencies, I assess its prospects for effectiveness and consider its feasibility in the current urban political-economic context.

EXPLORING VISION AND SUBSTANCE

Vision

At the broadest level, the vision projected by the community-based economic development strategy (or what Soifer, 1990, p. 240, refers to as the "new community economics movement") embodies an image of a new, radically different mode of political-economic organization for society. This new mode of political-economic organization represents a so-called third way—a system departing in meaningful ways from the two extant forms of modern political-economic organization: corporate capitalism and bureaucratic socialism (see Brown, 1993, p. 205; Dahl, 1985). According to this ideal, reorienting the nature of ownership and control of productive assets in the economic development process to favor the increased power of nonprofit community-based institutions becomes the constructive foundation of, as one author writes, "a new system . . . capable of going beyond the capitalist market and the regulatory state as characterizing features of political economy" (Bruyn, 1987, p. 6).[2]

In addition to being a blueprint for broad political-economic reconstruction, the vision projected by this strategy takes on other dimensions. Most obviously, the vision is in some sense "communitarian": The endeavor of community-based economic development—as once again, understood as an ideal—downplays excessively individualistic conceptions of well-being and achievement and, instead, accentuates the notion of human beings expressing a shared commitment to common ends (Kelly, 1977, pp. 135-136; Wiewel et al., 1993, p. 96; also see Sandel, 1982). Furthermore, in light of the communitarian ethic pervading its institutional structure, this strategy envisions an economy based to a greater degree on the concept of "mutual aid" and, more generally, envisions a society where cooperative, voluntaristic forms of interaction increasingly become a dominant feature of human behavior (Bruyn, 1987, p. 17).

The ideal of democratizing control of local economic resources in a direct, participatory manner provides a third major theme in the vision of the community-based strategy (Kelly, 1977, pp. 142-154; also see Pecorella, 1985). Rooted in the "citizen control" movement of the 1960s (see Arnstein, 1969), this strategy envisions lodging the authority to make key decisions concerning economic development matters in the hands of an active and informed local citizenry. This "participatory democracy" ideal evokes not only an institutional structure for citizen self-determination; it also is seen both as a basis for the fuller individual self-development of creative human capacities and, more specifically, as a basis for persons to acquire and exercise the skills of citizenship (see Barber, 1984; Macpherson, 1977; Pateman, 1970).

Three final elements of this vision bring it into sharper focus. First, as implied by both its "communitarian" and "participatory democracy" ideals, a strong measure of decentralization is imagined by the community-based strategy.[3] The conceived jurisdictional scope of many of the key institutions for community-based economic development is usually quite narrow, as its organizations are often based at the neighborhood level (see Perry, 1987; and more generally, Kotler, 1969; Morris & Hess, 1975). Second, this strategy sees the actions undertaken by community-based economic developers as part of a larger grass-roots protest effort initiated by those citizens, often but not always members of minority groups who have been marginalized by traditional economic institutions (see Boyte, 1980, chap. 5). Understanding local economic development as a protest activity in this way keeps a strong emphasis on political conflict as a means of achieving desired development goals.[4] Finally, community-based economic development is envisioned as an undertaking based on the principle of "self-help" (Giloth, 1988, p. 343): Through the mechanisms afforded by community-based institutions, the conception at work here is that residents of distressed urban communities themselves will be able to take responsibility for improving their economic and social conditions.

Substance

How is the community-based strategy substantively constituted? That is, how would a city reorient its economic development efforts in a community-based direction? As we shall see, much more than the two other strategies, this approach emphasizes the building of new economic institutions for local development—which are in turn nurtured and buttressed by cities through some combination of financial and technical assistance.[5]

A model of key community-oriented enterprises developed by Severyn T. Bruyn adeptly illustrates the essential nature and aspirations of these institutions, as well as their relationship to one another. The basis of this model derives from its specification of the two fundamental goals of community-based economic development: increasing "local autonomy" and strengthening "local economic viability" in the economic development process. The former refers to the degree of control local people have over their economic resources; the latter to the degree local people can successfully develop these resources (Bruyn, 1987, p. 8). According to this model, achieving local autonomy entails the transformation of each of the three key factors of production—land, labor, and capital—in ways giving each a community base. Strengthening local economic viability requires the development of community-based consumer corporations, as well as a community-based apparatus to plan, coordinate, and manage the local economic development process (Bruyn, 1987, pp. 8-18). Below, I discuss the specific characteristics of the individual institutions of this model.

Increasing Local Autonomy: Transforming Key Factors of Production: Land, Labor, and Capital. The first factor of production, *land*, is brought under community control through an institution known as a community land trust (see Bechtel, 1989, p. 18; Davis, 1983; Gunn & Gunn, 1991, pp. 94-97; Soifer, 1990). As Kirby White and Charles Matthei (1987) explain, a

TABLE 4.1   Elements of the Community-Based Strategy

Community Land Trusts

Description:	Holds land in trust for community benefit
Benefits:	Community control of land, less absentee ownership and speculation, fairer allocation of equity
Key uses:	Housing, economic development

Worker-Owned Firms

Description:	Employee ownership and democratic self-management of business firms
Benefits:	Higher worker incomes, anchors capital in communities
Key types:	Producer co-ops, ESOPs

Community Finance Institutions

Description:	Community-based banking/finance
Benefits:	Provides capital for community economic development
Key types:	Community development credit unions, community loan funds, community development banks

Consumer Co-ops

Description:	Social ownership of consumption means
Benefits:	Supplies goods/services at reduced cost, surpluses (profits) returned to community
Key types:	Food, housing, health care

Community Development Corporations

Description:	Management of community economic development process
Benefits:	Comprehensive planning, coordination
Key projects:	Business development, commercial real estate, housing

community land trust is "a democratically structured not-for-profit corporation, with an open membership, created to hold land for the benefit of a community and of individuals within the community" (p. 41). Governed by elected boards typically composed of users of the land, other community residents, and public interest representatives, community land trusts "provide the means of holding legal title and rights to property in a social, rather than individual or business-directed, corporate form" (Gunn & Gunn, 1991, p. 94).[6]

The community land trust acquires land through either purchase or donation with the intention that it will be held in permanent steward-ship by the community trust. Although the trust retains ownership, land is made available to individuals, businesses, and other organizations for housing and/or economic development purposes through long-term leases. Because the equity of these leaseholders is limited to the amount of their actual investments (purchase price and the cost of major improvements, adjusted for inflation and depreciation), the land trust "distinguishes between that portion of the real estate's market value that is created by the individual home or business owner through investment of labor or capital, and that part that is created by the community or the general public through community development or improvement efforts, public investment, or broader economic forces" (White & Matthei, 1987, pp. 41-42). Hence this legal separation between the community ownership of the land and the individual ownership of the property occupying that land introduced by the community land trust allows this institutional innovation to allocate "equity fairly according to its source, and value created by the community is retained for the public good" (White & Matthei, 1987, p. 42; see also Soifer, 1990).[7]

Because the control of land is so critical to any successful economic development strategy—especially in the urban context (see Garber, 1990; Logan & Molotch, 1987)—the community land trust, which furnishes the local community with that control, plays a central role in the community-based strategy for alternative local economic development. Moreover, beyond providing for this general community control of land —which also in turn limits the destructive effects of absentee ownership and speculation (see Gunn & Gunn, 1991, p. 95; White & Matthei, 1987, pp. 50-51)—the community land trust also promotes community-based economic development more narrowly by allocating that equity not created by private actors back to the community as a whole.

Labor as a factor of production achieves a community orientation by organizing the business firm as a producer (worker) cooperative (Bruyn, 1987, pp. 10-11) or, conceivably, as an employee stock ownership plan (ESOP). Although these organizational forms differ significantly from one another (see Ellerman, 1987), both allow for, in varying degrees, worker ownership and democratic employee self-management of the company. Thus, through these institutional designs, economic assets are more broadly based in the community, as their ownership and control is extended beyond both individual entrepreneurship and the traditional business corporation (see Alperovitz, 1991).[8]

One economic development benefit from employee ownership is that workers in the community, as owners, would likely see their incomes rise as they divide and share the company's surplus (Dahl,

1985, pp. 105-106). Yet for our purposes, a point made by Olson (1987) suggests an even more crucial benefit: As an economic development tool, employee ownership helps anchor capital in local communities.

Financial inducements often entice owners of conventional companies to move their operations or to invest the company's capital base in other supposedly more profitable enterprises. In contrast, employee owners have a strong incentive to ensure both that their enterprises continue to operate in their communities and that accumulated capital is reinvested in ways fostering the company's long-term market competitiveness (see also Dahl, 1985, pp. 120-122). In general, then, when workers are owners, their own interests in maintaining stable employment and job security accord with a local community's interest in preventing erosion of its tax base and unemployment among its citizens. Moreover, even partial employee ownership can provide workers with considerable leverage in decisions on capital disinvestment and company mobility (Olson, 1987, pp. 244-247).

As specified in the Bruyn model of community-based enterprises, community finance institutions supply the final, and most crucial, factor of production—*capital*—for local economic development. These financial institutions have three distinct community-based institutional forms: the community development credit union, the community loan fund, and the community development bank.

A community development credit union, like all credit unions, is "a cooperative, nonprofit corporation created by and for people affiliated by a common bond, for the purposes of promoting thrift among its members and of loaning funds to its members at reasonable rates" (Swack, 1987, p. 90; also see Boyte, 1980, pp. 136-137; Curtis et al., 1985, p. 97; Gunn & Gunn, 1991, pp. 62-66).[9] The "common bond" in the case of community development credit unions is residence: They are designed to serve the financial needs of the communities in which they are based (Swack, 1987, p. 91). Essentially a customer-owned local bank (Bruyn, 1987, p. 12), the community development credit union is organized democratically according to the one-member, one-vote principle. The organization's members, who are residents of the local community, elect the board of directors and set the credit union's agenda (Swack, 1987, p. 92). This agenda revolves around the goal of using the funds deposited in the credit union to mobilize capital for community revitalization, in essence ensuring the reinvestment of local residents' personal savings back into their communities (Rosenthal & Washington, 1987, p. 190; Wiewel & Weintraub, 1990, p. 168).

Community loan funds act as financial intermediaries between investors in community economic development and individual community-based developers. They accept loans from a variety of sources and use this capital to provide loans for community-based development projects (Swack, 1987, pp. 82-84). Funds can be capitalized by the retained earnings of community-based enterprises or by money from a variety of entities, including charitable foundations, religious groups, individuals, and corporations (Gunn & Gunn, 1991, p. 67), or conceivably— and most important for our purposes—a city administration working to reorient its economic development strategy in a community-based direction.[10] Another important potential source of capital for these community-based loan funds is the Community Reinvestment Act, which requires conventional banks to meet the credit needs of their communities (see Shlay, 1989, pp. 202-204). Many local community groups have used the act to challenge the "redlining" lending practices of banks, cajoling bankers to make future commitments to invest in low- and moderate-income neighborhoods. With some implementation changes, such money could be channeled through a community loan fund.[11]

Community development banks are in many ways like traditional banks, save they have a strong community orientation and are designed especially to aid in the economic revitalization of poorer urban neighborhoods (Bruyn, 1987, p. 13; Gunn & Gunn, 1991, p. 69; Parzen & Kieschnick, 1992).[12] Investors in community development banks would include local residents and many of the groups mentioned above in regard to community loan funds, as well as cities trying to promote community-based economic development in their jurisdictions.[13]

Strengthening "Local Viability:" Creating Consumer Corporations and an Apparatus to Plan, Coordinate, and Manage Community-Based Economic Development. The most important form of consumer corporation in the community-based strategy (as outlined here) is the *consumer cooperative (co-op)*. "The consumer cooperative," as Jean Hammond (1987) explains, "is a self-help structure that works directly on the social and economic problems facing many communities today" (p. 98).[15] Controlled democratically according to the "one-member, one vote" principle, these

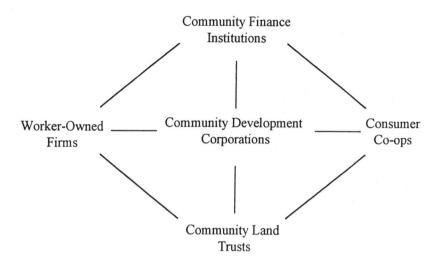

Figure 4.1. Model of the Community-Based Strategy for Alternative Economic Development
SOURCE: Developed, with relevant modifications, from Bruyn (1987, p. 17).[14]

organizations "provide high quality consumer goods and services to their members at a reasonable price" (Gunn & Gunn, 1991, p. 99). Any surplus generated by the consumer co-op is returned to its members (Hammond, 1987, p. 99). From an economic development perspective, then, the formation of a community-based consumer co-op provides urban residents with an institutional vehicle to receive necessary goods and services at a reduced cost, increasing the net wealth of the community.[16]

At the center of the Bruyn model is the *community development corporation (CDC).*[17] This entity, according to Bruyn (1987, p. 16), "is a democratic firm designed to be accountable to *all* residents of the community. . . . Residents become members for a small fee and may participate equally with one vote in shaping community policy" (p. 16). As the model's comprehensive planning and management vehicle, CDCs have "an overall responsibility to help develop land, labor, capital." They are "the crucial coordinating agent" in the community-based economic development process, ensuring that the activities of land trusts, self-managed firms, and community finance institutions work toward common developmental objectives (Bruyn, 1987).

In more specific terms, CDCs are nonprofit, usually tax-exempt organizations, although some have for-profit subsidiaries. Their roots can be traced to the Great Society era of the 1960s, yet most CDCs operating today have come into being since then (Wiewel & Weintraub, 1990, p. 160). Existing CDCs exhibit incredible diversity in terms of size and mix of programmatic activities (Roberts, 1980, pp. 32-34); many engage in activities outside the economic development arena, including social service delivery, cultural activities, job training and placement, the promotion of community pride, and advocacy (Vidal, 1992, p. 5; Zdenek, 1987, p. 115). Nevertheless, most CDCs focus on achieving some form of economic development in their jurisdictions (see Vidal, 1992, p. 5). They are governed by boards of directors composed primarily of people with a strong stake in the community, many of whom are residents of that community (Vidal, 1992, p. 39; also see Kelly, 1977). All CDCs focus their energies in a clearly defined geographic area, which can be citywide or neighborhood specific, and always includes a high concentration of low-income persons (Giloth, 1988, p. 344; Peirce & Steinbach, 1987, p. 13).[18]

ASSESSING PROSPECTS FOR EFFECTIVENESS

To begin the evaluation of this second strategy's potential to bring about the necessary reduction in the structural dual dependencies, we consider its prospects for effectiveness by marshaling a variety of empirical evidence.

Overarching Issues

Economic Theory

Earlier I pointed out that, as "alternative" approaches to urban economic development, each of the three strategies challenges the prevailing economic orthodoxy (neoclassical theory). In the case of the community-based economic development, this challenge is directed at the individualism (or "atomism") so central to that orthodoxy. Specifically, this strategy is based on the theoretical premise that successful accumulation need not necessarily be done atomistically, but instead can be accomplished in *collective* fashion.

Two general arguments support the plausibility of this claim. First, a key element of the economic development process—entrepreneurship (or what Schumpeter, 1942/1976, called the "entrepreneurial function") —is already largely a *social* effort. As Meehan (1987) writes,

> To a great extent the entrepreneurial function is already socialized; however romantically we view computer inventors working in their garages, they are the exception to the rule in a sector dominated by the research activities of giant corporations and government agencies. . . . The question is not whether entrepreneurship should be individual or socialized, but *how* it should be socialized. (p. 135)[19]

In short, the collectivism of community-based economic development is simply one example of how the socialization of entrepreneurship can occur (cf. Polsky, 1988, p. 14). Therefore, this strategy proposes nothing that isn't already at work in the development process.

The prosperous experiences of "immigrant economic development" from the late 19th century onward provide a second argument supporting the plausibility of successful collective accumulation. Historically, as Cummings (1980) writes, "Many immigrant groups [in the United States] responded *collectively* to the conditions posed by urban and industrial life. The self-help institutions created by many ethnic minorities emphasized collective rather than individualistic modes of development" (p. 6; also see Butler, 1991, pp. 22-26). The key to the economic achievement of individuals in these groups was their cooperation with others and, more generally, the supportive nature of the community environment (Kelly, 1977, pp. 15-16, 135-136). As a result, public officials should learn from this history of success, and understand that the collectivism of the community-based strategy similarly can serve as "an effective way to mobilize and distribute resources" (Cummings, 1980, pp. 25-26) for economic development.

The utility of the collective approach to economic development is especially clear in the poorer areas of the city, which are the major focus of community-based development efforts. In this setting, collectivism has special advantages over the individualistic (or atomistic) approach enshrined in orthodox economic theory (see Harrison, 1974, pp. 165-167). Individuals acting without any collectivist support system in an economically distressed area such as an urban ghetto face significant obstacles to success (Faux, 1971, pp. 45-47; Kelly, 1977, pp. 17-18). Moreover, when the individuals do succeed, they tend to exit from the poor area, taking their capital—physical as well as human—with them and leaving little behind for the remaining community (Fusfeld & Bates, 1984, chap. 10; Osborne, 1988, p. 301).

An additional challenge to the neoclassical orthodoxy follows from this strategy's more general claim about the prospects for collective economic accumulation. This second challenge, a corollary of the broader indictment, imputes the economic orthodoxy's theoretical notions regarding human motivation. Namely, the idea of collective economic accumulation with its communitarian ethos implies a more complex conception of that motivation; in particular, it implies that the neoclassical conception based solely on material self-interest (narrowly understood) is fundamentally flawed (cf. Kelly, 1977, pp. 135-137). The community-based strategy does *not* require "a different sort of human being . . . the 'new man' of various political theories"—to use Stephen Elkin's (1987, p. 180) phrasing—that is, individuals need not be totally other-regarding in their behavior toward fellow members of the "community." But insofar as the neoclassical model of human motivation does hold, collective economic efforts are likely to face profound difficulties arising from the structure of material incentives, the most important being the infamous "free rider problem" (see Olson, 1982; Taylor, 1987).

The role that narrow, material self-interest plays in shaping human motivation has been the subject of intense academic debate and controversy. A large body of research points away from the neoclassical conception (see Mansbridge, 1990).[20] But perhaps the most perceptive (and commonsensical) approach to the question is to view it as a highly *contingent* matter and, as such, to understand that human motivation is a function of context. Most important, the structural design of institutions can have a strong effect on human behavior (cf. Elkin & Soltan, 1993). Researchers have found, for example, that decentralized institutional settings foster the development of cooperative rather than self-interested modes of association (Dryzek, 1987, pp. 221-225). This finding particularly strengthens the prospects that the community-based strategy will not necessarily fall victim to problems stemming from the inadequate provision of material incentives, as the conceived jurisdictional scope of many community-based economic institutions is quite narrow (e.g., many are designed to operate at the neighborhood level).

Legal Context

Aspects of the legal environment in which community-based economic development operates need not be extensively reviewed here: Unlike the other two strategies for alternative local economic development, nothing *inherent* in the community-based strategy raises troubling legal questions. The "collective" economic institutions of this strategy—land trusts, worker-owned companies, community finance institutions, and community development corporations—are all well-established in current legal doctrine. (See, generally, Bruyn & Meehan, 1987).[21]

Specific Issues

Moving on to more specific issues, we now investigate the prospects for effectiveness of the community-based strategy by looking at (a) the general possibilities for economic development in poorer urban communities and (b) the potential of the individual community-based institutions that constitute the model sketched above.

The Development Potential in Poor Urban Communities

To understand this strategy's prospects for effectiveness, we should first consider the general possibility for economic development in poorer urban communities. As noted above, the community-based strategy—unlike the other two strategies for alternative urban economic development—focuses special attention on lessening the uneven development of cities by strengthening the economic base in these poorer areas (see Peirce & Steinbach, 1987, p. 13; Vidal, 1992, p. 80). The collectivist approach of this strategy, as also noted above, may have distinct advantages over an atomistic approach in this setting. But the crucial question is whether development can be pursued successfully *at all* in areas of high economic distress, high crime, high incidence of drug and alcohol abuse, and a concentration of family dissolution (Shiffman, 1990, pp. 10-11; also see Wilson, 1987, pp. 21-29). Or is it simply the case that, as former Federal Reserve Board member Andrew Brimmer argued a quarter century ago, such investment would be socially wasteful and inefficient given the "unfavorable economic climate" in these areas (Brimmer & Terrell, 1969, quoted in Harrison, 1974, p. 164; see also, more recently, Lemann, 1991).

This is not to say that community-based developers cannot extend the model beyond these troubled communities; indeed, as we shall see below, this strategy's chance for success may be ultimately dependent on its extension to the relatively prosperous parts of the city, including the central business district.[22] The areas of the city targeted for community-based economic development may include not only its very poorest sections (i.e., "urban ghettos"), but also its moderate-income communities as well (see Wiewel & Mier, 1986, p. 205).[23] Nonetheless, the prospects for this strategy are intimately linked to the possibilities of creating viable economic activity where none currently exists (cf. Osborne, 1988, pp. 299-316).

Accurately evaluating these possibilities is a difficult and uncertain enterprise. These communities have suffered decades of structural disinvestment, which in turn constrains current economic activity (Betancur et al., 1991, p. 208; Bradford, Cincotta, Finney, Hallett, & McKnight, 1981, pp. 135-136). On the one hand, absent an initial round of stabilizing reinvestment of sufficient magnitude, determining the long-term market potential of these areas is impossible. On the other hand, pace Brimmer, market forces have so thoroughly neglected these communities that remaining incredulous about their potential seems quite reasonable.

Nevertheless, two pieces of evidence point to the possibility that these areas may have been hitherto irrationally overlooked by economic actors.

First, lenders required by the provisions of the Community Reinvestment Act to make home mortgages (and other loans) in lower-income neighborhoods have not found these investments to be appreciably riskier than loans in higher-income neighborhoods (Raven, 1991). In fact, bankers in New York (Alpert, 1991), Philadelphia (Wayne, 1992), and Milwaukee (Squires, 1992a) all report that money can be made from these activities (also see Hasell, 1994).[24] Moreover, a formal, detailed evaluation of reinvestment act programs in Chicago concluded that the "performance of reinvestment loans has been quite good—with some of the major reinvestment programs reporting losses that are much lower than their regular portfolios" (Bradford & Cincotta, 1992, pp. 263-264, 278-279). Likewise, in a program in Pittsburgh, none of the 272 residential mortgage loans made by Union National Bank under a Community Reinvestment Act lending agreement had defaulted as the program began its third year (Metzger, 1992, p. 89; Squires, 1992a, p. 23).

Second, other analyses demonstrate the strong potential for successful business development in the inner city, despite the traditional private sector's current lack of interest. Recent marketing research, for example, shows that the internal market for goods and services in poorer inner cities—especially in the retail sector—is surprisingly favorable for profitability, despite the concentration of lower-income citizens in these areas. This phenomenon arises because of an acute lack of competition in the inner city—where consumers remain "gravely underserved" by local retailers—while, at the same time, the suburban retail markets have of late become "saturated" (Alpert, 1991, p. 168).[25] And in regard to the development of businesses serving the larger regional economy, the inner city often has a distinct comparative advantage over other areas based on location (Porter, 1995).[26]

Individual Institutions

Even if an elementary development potential exists in poorer areas of the city, the more important question is whether new community-based economic institutions work sufficiently well to exploit this potential. Below, I discuss the prospects for effectiveness of each of the four major institutional forms.

Community Development Corporations. The CDC, as noted previously, lies at the heart of the Bruyn model for community-oriented enterprises. It is the model's "central coordinating agent" with the overall responsibility to plan and manage the development of each of the three key factors of production (Bruyn, 1987, pp. 16-17; Giloth, 1988, p. 348). Most CDCs currently are developed only to a rudimentary degree and hence are not yet functioning in this manner (Soifer, 1990, p. 240). Nevertheless, to understand the future prospects for effectiveness of this crucial institutional form, the past performance of CDCs as a community-based economic developer must be assessed.

First, consider some basic facts concerning the pervasiveness and output of this institutional form: In the mid-1970s, there were about 200 CDCs in the United States; by 1988, a survey undertaken by the National Congress for Community Economic Development put the number in the 1,500 to 2,000 range, four fifths of which were found in urban areas (1989).[27] By 1988, CDCs had (a) "built nearly 125,000 units of housing," (b) "developed 16.4 million square feet of retail space, offices, industrial parks and other industrial developments," (c) "made loans to 2,048 enterprises, equity investments in 218 ventures and own and operate 427 businesses," and (d) "accounted for [the] creation and retention of close to 90,000 jobs in the last five years" (NCCED, 1989, pp. 1-2; cf. Vidal, 1992, pp. 85-99).

Thus, CDCs are no doubt nearly omnipresent in urban America, and their economic development outputs are clearly impressive. As a common and active vehicle for urban economic development, then, CDCs have established a solid base from which to build. Yet to more accurately assess their future potential, we must examine not merely what they have done but also how well they have done it.

Unfortunately, despite the pervasiveness of CDCs, there are few comprehensive evaluation studies of their performance (Giloth, 1988, p. 348). However, the results of those few studies that have been conducted are generally positive (see Perry, 1987). In summary, CDCs were shown to be "able to complete most of the tasks they set out for themselves (such as financing, acquisition, construction activities, planning, and obtaining funding)" (Wiewel & Weintraub, 1990).

But because much of the current economic development efforts of CDCs centers on their actions as nonprofit developers of affordable housing (see Goetz, 1992; Mayer, 1984), the fact that CDCs were able to complete most of their tasks reveals little about their future potential for strengthening the capacity for genuine community-based accumulation. This is because *business ownership* (either whole or partial) is the CDC activity most important for bringing about the realization of that potential (cf. Alperovitz, 1991, pp. 1-2). Wiewel and Mier (1986) emphasize this point: "The business activities [done] by community organizations could be useful for changing [the local economy], eventually bringing a greater proportion [of it] under community control" (p. 223).

Yet as the existing evaluation studies also show, this CDC task has proved to be most difficult to accomplish (Wiewel & Weintraub, 1990,

p. 173; also see Vidal, 1992, p. 79). Nevertheless, even in this area, the evaluative evidence is encouraging. For example, in the first extensive evaluation study of CDCs, a private research firm, Abt Associates, uncovered evidence pointing in a positive direction: Analyzing numerous economic development projects undertaken by 30 CDCs over a 3-year period (1970-1973), it projected that CDC business ventures "would have no more than a 50 percent failure rate; that is, 50 percent of CDC ventures would at least be breaking even after four years" (Perry, 1987, p. 187). Given the high failure rate of private-sector small business—generally thought to approach 80%—this statistic is impressive (Perry, 1987, p. 187; see also Giloth, 1988, p. 348; Wiewel & Weintraub, 1990, p. 173).

A systematic analysis of 15 CDCs that operated sometime during the period from 1968 to 1980 later confirmed this projected result. This study, undertaken by the National Center for Economic Alternatives, also found that "ventures had a business survival rate of about 50 percent" (Perry, 1987, p. 192). Even though the study noted that "CDCs tolerated a level of loss in ventures that would not conventionally be accepted" (p. 192), the data on losses and gains for the 288 CDC business ventures analyzed showed that, in aggregate, these endeavors netted a small profit. Likewise, a study of 17 such enterprises in Chicago yielded results broadly consistent with the studies by Abt and the National Center for Economic Alternatives; it found a lower number of ventures to be profitable, but the 20% private sector benchmark was in fact met (Wiewel & Mier, 1986, p. 219).

Despite these somewhat positive evaluative results, the overall economic development record of CDCs—in business development as well as in other arenas—clearly "has not lived up to its promise" (Daniels et al., 1981, p. 179), has been generally "uneven" (Levine, 1989, p. 30), and has spawned many economic development projects that can be described only as "disasters" (Twelvetrees, 1989, p. 157).

Yet there is reason to believe that, as the CDC movement matures, this record will improve: One firm conclusion reached by numerous studies is that failures of CDC projects are often attributable to lack of management and organizational capacity (Daniels et al., 1981, p. 181; Giloth, 1988, p. 348; Perry, 1987; Wiewel & Weintraub, 1990, p. 174).[28] Today's CDCs are, however, rapidly building these management skills and a more general organizational capacity to act, accomplishing complex economic development projects requiring great expertise and skill (Peirce & Steinbach, 1987, p. 30). Moreover, research has isolated the crucial dynamic at work, showing that the "process of engaging in economic development activities strengthens the CDCs as organizations and thereby increases their capacity to do additional development work" (Vidal, 1992, p. 99). In essence, the CDC movement—which in the 1960s and 1970s had plentiful financial resources but little managerial/ organizational capacity—now has plentiful capacity but few financial resources (Peirce & Steinbach, 1987, p. 32). Hence, reorienting a city's economic development funding in a community-based direction is, under present and future circumstances, now more likely to be an efficacious means by which to stimulate local economic vitality.[29]

Community Finance Institutions. The community-based strategy envisions that new community finance institutions will be the key supplier of the most critical factor of production, capital, to the local development process.

The movement to create these institutions—community loan funds, community development credit unions, and community development banks—was recently bolstered by the nationwide excitement generated by the successes of the South Shore Bank, a community development bank in south Chicago. The South Shore Bank is the centerpiece of a "multipronged development instrument" (Taub, 1988, p. 5) known as the Shorebank Corporation (formerly Illinois Neighborhood Development Corporation), a for-profit bank holding company with three subsidiaries apart from the bank itself. This community development bank has, in short, been a successful supplier of investment capital to fund the revitalization of its surrounding neighborhood (see Gunn & Gunn, 1991, pp. 70-72; Osborne, 1988, pp. 305-308; Parzen & Kieschnick, 1992; Taub, 1988). The bank showed annual profits of more than $1 million, and its assets grew from $40 million in the mid-1970s to $135 million in the late 1980s. With more than $140 million invested in the South Shore neighborhood, by 1988 the bank had financed the rehabilitation of more than 8,000 housing units—about one quarter of the entire housing stock in the neighborhood (Peirce & Steinbach, 1987, p. 68).

The recent restructuring of the traditional banking system appears to be buttressing prospects for the further development of economically successful community-based financial institutions. Changes in the domestic regulatory environment in which commercial banks operate are spurring a new era in banking, one marked by increasing industry consolidation and concentration (largely through merger and acquisition) and the resulting rise of so-called megabanks. A restructured system, many believe, will serve to strengthen the ailing industry through the increased efficiency of banking operations. Yet because these megabanks will compete nationally and internationally, they likely will be less interested in local economic development—a traditional concern of the smaller banks they are replacing. The traditional banking system's focus on corporate and international markets (cf. Bradford & Cincotta, 1992, pp. 264-265; Squires, 1992a, pp. 27-28) leaves open a potentially profitable domain for the new breed of community-based financial institutions emerging in America's cities.

This probable market disengagement by megabanks, it should be noted, is not just the familiar aversion the financial community has shown to development activities in the poorest areas of the city. Rather, it extends beyond these older redlining practices, and would include a lack of interest on the part of commercial banks in all but the most appealing local projects. Moreover, development lending in cities would not be the only market ceded by these new consolidated and concentrated entities. Retail services might be affected as well. In response to its past financial problems, the banking industry has focused on the profitability of each operating unit within the larger corporation. When acquisitions and mergers occur, retail offices and branches with lower profit margins compare unfavorably to the several recently acquired branches in high-growth areas. As a result, consolidation often brings about the closing of these lower-profit operations (Bradford & Cincotta, 1992, p. 265). In this setting, new community-based financial institutions (such as community development credit unions) can step in and fill the void, further increasing the amount of economic activity under community control (Rosenthal & Washington, 1987, pp. 184-192). And evidence from New York City suggests that this phenomenon is beginning to take root. Several community-based credit unions have spouted in poorer neighborhoods, sometimes operating in the very buildings left behind by traditional banks (Purdy & Sexton, 1995).

Worker-Owned Companies. Labor as a factor of production becomes community-based when workers obtain ownership and control (self-

management) of their companies (cf. Bruyn, 1987, pp. 10-11). Two institutional vehicles accomplish this transformation: the producer (worker)
cooperative (co-op) and the employee stock ownership plan (ESOP).
Co-ops generally provide a purer form of worker ownership and control because only workers are owners of the company and internal
decision making is based on the one-person/one-vote principle (cf. Ellerman, 1987, pp. 234-237; Dahl, 1985, pp. 148-150). Nevertheless, properly structured ESOPs (i.e., those with majority stock ownership by
employees and democratic decision-making procedures) also base labor as a factor of production in the local community (cf. Swaine, 1993;
Shavelson, 1990, pp. i-ii).[30]

To assess the future effectiveness of worker ownership, we must first
evaluate the past performance of these types of companies. At a general
level, evaluative research on this question suggests that worker-owned
enterprises perform as well or better than conventional business firms
in such areas as profitability, productivity, job generation, and growth
(see Dahl, 1985, pp. 120-133).

For example, numerous studies comparing companies with ESOPs
to those without ESOPs reach this conclusion. In a survey of 229 ESOP
companies, Marsh and McAllister (1981) found that "companies with
ESOPs had average annual productivity increases 1.5 [%] greater than
the national production average for the period 1975 to 1979" (cited in
Rosen & Wilson, 1987, p. 212). A survey of 98 companies by Rosen and
Klein (1983) estimated annual employment growth for ESOP companies to be 2.78% higher than non-ESOP companies, "resulting in a 31
percent increase in jobs over 10 years" (Swaine, 1993, p. 302). Conte and
Tannenbaum (1978) also surveyed 98 employee-owned companies;
they found these companies to be "on average 1.7 to 1.5 times more
profitable than the non-employee-owned companies in the same industry" (Swaine, 1993, p. 302). Finally, Quarrey (1986) collected data on 55
employee-owned companies over a period of 15 years and found that,
compared with similar companies lacking an ESOP, companies with
ESOPs grew 3.84% quicker per year in terms of employment level and
3.51% quicker in their sales (Rosen & Wilson, 1987, p. 220). The one
anomalous study, conducted by the U.S. General Accounting Office
(1987), investigated 111 employee-owned companies and found no
relationship between the establishment of an ESOP and an improvement in profitability and productivity. This research has been criticized,

however, as lacking methodological rigor, and hence, its conclusions are suspect (Swaine, 1993, pp. 303-304).[31]

Further illustrating the strong economic performance of worker-owned companies are the "stunning successes" of two groups of co-ops: the Mondragon cooperatives in Spain and the plywood cooperatives in the Pacific Northwest region of the United States (Dahl, 1985, p. 131). By the mid-1980s many companies in the latter group had been in existence for 25 to 30 years, making the plywood co-ops among the most enduring examples of worker-ownership and employee self-management in the United States. Compared with conventional companies in the industry, the co-ops exhibited spectacular prosperity and a high level of productivity (Dahl, 1985, p. 140; Greenberg, 1986, pp. 28-29).[32] The history of the Mondragon experience is even more impressive. Building from a single co-op organized in 1954, this complex of enterprises grew to include more than 85 cooperative companies employing some 20,000 people, increasing their market share and sales at a dramatic rate (Dahl, 1985, pp. 123-124; Shavelson, 1990, p. 129).[33] Equally stunning is Mondragon's job creation record. Even between the years of 1977 and 1981—a period of economic recession and rising unemployment in the rest of the Spanish economy—employment in the Mondragon co-ops grew from 15,700 to almost 18,500 (Dahl, 1985, p. 132).

The body of evidence presented thus far is fairly extensive and compelling. Yet evaluating the effectiveness of Bruyn's (1987) notion of basing labor as a factor of production in the community (via worker ownership and control) turns, crucially, on the proposition that *worker participation in workplace decision making* fails to harm—and actually improves—the productivity and general performance of a company. The logic behind such a proposition is that worker-controlled companies are "likely to tap the creativity, energies, and loyalties of workers" (Dahl, 1985, p. 132) to an extent far exceeding the ability of conventional companies to do so.

Researching this proposition, Dahl (1985) concludes that a "broad range of experiences" show that "participation by workers in decision-making rarely leads to a decline of productivity; far more often it either has no effect or results in an increase in productivity" (p. 33). Moreover, the U.S. General Accounting Office (1987) study referenced above, though not finding an increase in productivity growth for all ESOPs, did discover a "clear upturn in productivity growth for those . . . with a high degree of employee participation" (Swaine, 1993, p. 303). Likewise, Quarrey's (1986) analysis (also referenced above) found that although all the ESOP companies, taken together as a group, outperformed non-ESOPs, those that were the most participatory improved their economic performance at an annual rate about three times greater than the strong average performance registered by all ESOPs (Rosen & Wilson, 1987, p. 220).

Community Land Trusts. Land, the final factor of production, is brought under community control by community land trusts (cf. Bruyn, 1987, p. 10). Compared to the other individual institutions of the community-based strategy, we know relatively little about these trusts. They are "relatively young and small" (White & Matthei, 1987, p. 54). Of the 65 or so around the country by the late 1980s, many had been in operation for less than 5 years (Soifer, 1990, p. 240).

At least one in-depth analysis of a community land trust has been conducted to date. This study, Soifer's (1990) look at the well-known Burlington (Vermont) Community Land Trust, focused on the ability of this institutional form to create affordable home ownership for low- and moderate-income individuals—a key goal of the trusts. Under the trusts, home ownership becomes more affordable because the cost of the land often amounts to 20% or 25% of the purchase price (Soifer, 1990, p. 238). On this score, the Burlington Community Land Trust, created in 1984 by the progressive city government under Mayor Bernard Sanders, helped some moderate-income persons buy homes "they otherwise would be unable to afford because of tight market conditions" (Soifer, 1990, p. 249; White & Matthei, 1987, p. 47).

Nevertheless, Soifer (1990) admits that "the BCLT (Burlington Community Land Trust), at best, must be recognized as a partial solution to the affordable housing crisis in Burlington" (p. 249). The trust failed to make home ownership attainable for low-income people, for example. They remain priced out of the city's housing market. But those living as renters in properties owned and managed by the trust found that the organization provided them with "a somewhat more beneficent landlord" (Soifer, 1990, p. 249).[34]

The key to the future success of the Burlington Community Land Trust is growth (Soifer, 1990, p. 247). The trust does plan to make a major push for expansion; by the year 2004 it hopes "to own 25% of the residential property in Burlington." Moreover, the land trust plans to expand its scope by including nonprofit businesses and other nonprofit organizations providing services to the community as occupants of about 10 percent of its properties (Soifer, 1990, p. 244). Yet in his assessment of the prospect that the trust can actually own 25% of the city's residential property, Soifer argues that such a goal "appears unrealistic. Nevertheless," he continues, "it [does] demonstrate the BCLT's commitment to becoming a major player in the city's real estate market" (Soifer, 1990, p. 244). And it is clear that initial growth of a community land trust can fuel even more growth: "As we acquire properties," an official of the trust said, "the corporation [the Burlington trust] is acquiring assets, against which [we] can . . . borrow . . . to get more money" (quoted in Soifer, 1990, p. 246).

UNDERSTANDING CONSTRAINTS ON FEASIBILITY

The preceding discussion shows that the community-based strategy's prospects for effectiveness, like those of the entrepreneurial-mercantilist strategy, are encouraging. But as I said in Chapter 2, a strategy's potential to be the vehicle to diminish the structural dual dependencies at work in cities is also linked intimately with a strategy's *feasibility* in the current urban political-economic context. Can this strategy be implemented on a scope necessary to alter the prevailing structural arrangements in central cities?

Like the entrepreneurial-mercantilist strategy (and municipal-enterprise strategy discussed below), the community-based strategy, as a full-blown model of alternative urban economic development, also can be described as experimental in nature—with its empirical manifestation largely sporadic and embryonic. Hence, here, too, no example exists of a *comprehensive* development strategy consistent with the community-based approach (as best expressed by the Bruyn model sketched above).

Nevertheless, numerous cities have had extensive experience with CDCs—which Perry (1987) calls "the core institution" of community-based economic development. These institutions are mostly rudimentary in form (cf. Levine, 1989, pp. 30-31; Soifer, 1990, p. 240) and, as such, are not playing the central role in the local development process envisioned for them by Bruyn. Yet as Bruyn (1987) notes, perceptively, the hundreds "seeded" all over the country are in fact "part of the foundation for a new local economic order" (p. 16). More to the point, because CDCs are in fact the "central" or "core" institution of community-based economic development, it seems logical that they must be developed first if the broader strategy is to flourish.

One American city with considerable CDC activity is Pittsburgh.[35] As we will see below, the overall impact of these well-established efforts remains limited. Yet as with our scrutiny of St. Paul's experience with entrepreneurial mercantilism, an analysis of the Pittsburgh experience with the community-based strategy will contribute to our overall understanding of whether this second strategy can be feasibly implemented in contemporary central cities.[36] I accomplish this task by highlighting the nature of the constraints working to limit the strategy's effects in an empirical setting.

The Context: The Pittsburgh Post-War Regime

For most of the Post-War era, the political life of Pittsburgh has stood out as a nearly pure example of what has been identified analytically in this research as the "dominant urban regime form" (see Chapter 1): With the exception of one 7-year period of "interlude" (Stewman & Tarr, 1982, pp. 78-94), its regime has been constituted by (a) a "governing coalition" that has at its center a close alliance between officials of the local state and local corporate business interests and that (b) has actively pursued a policy agenda marked by a strong orientation toward the promotion of economic growth via corporate-center/mainstream strategies (see Coleman, 1983; Elkin, 1985; Jezierski, 1988, 1990; Lubove, 1969; Sbragia, 1989; Stewman & Tarr, 1982). The Pittsburgh regime, constituted as such, in many ways resembles the regime in place in post-War Atlanta (Stone, 1989, p. 236). Certainly both have been remarkably stable and enduring over the past 40 years.

The strong political alliance in Pittsburgh between public officials and corporate elites (often explicitly referred to as a "public-private partnership" in urban governance) dates back to the 1945 election of David Lawrence as mayor. Lawrence, the powerful boss of the city's Democratic political machine, and Richard King Mellon of Mellon Bank—the city's most important businessman—were the key players in the formation of this alliance. Two years earlier, Mellon had founded the Allegheny Conference on Community Development, a planning organization made up of the chief executive officers of major local corporations and endowed with considerable technical and professional skills. Together, the Allegheny Conference, as it is called, and Pittsburgh's public sector leaders, acting largely through the newly created, semiautonomous Urban Redevelopment Authority, launched a series of aggressive redevelopment efforts collectively known as Renaissance I (1945-1970) and later, under more complex but similar arrangements, Renaissance II (1977-1987) (Jezierski, 1990; Lubove, 1969; Stewman & Tarr, 1982; see also Sbragia, 1989, 1990).

At the heart of the urban development efforts undertaken by the Pittsburgh governing coalition or "partnership" during Renaissance I and II was the traditional emphasis on rearranging urban land use patterns to spur investment within the central business district. In response to economic restructuring, however, the partnership has refocused its development strategy over the past decade. The goal of this new strategy (Renaissance III) is to foster a transformation of the regional economic base through the development of new advanced technology industries (Ahlbrandt & Weaver, 1987; Sbragia, 1990). As a result of this shift in strategy, the segment of private sector elites playing major roles in the governing coalition has become less narrow. The Allegheny Conference and its allies in local corporate foundations still play a key role in the coalition, but influence is increasingly wielded by such coalition members as the Pittsburgh High Technology Council—a trade association of executives dedicated to the support of the local high-tech sector—and nonprofit institutional actors directly associated with advanced technology (such as Pittsburgh's two major research universities) (Ahlbrandt & Weaver, 1987; Jezierski, 1990).

Community-Based Development in Pittsburgh

Overview. Community-based economic development activity in Pittsburgh centers on three institutional actors: (a) a network of 10 "core" territorially based development organizations, (b) the Pittsburgh Partnership for Neighborhood Development (PPND), and (c) the city government.

The 10 CDCs vary in size and scope of activities, yet many are active in all three of the major arenas of CDC economic development: housing, commercial and industrial property development, and business development (see Vidal, 1992). Although neighborhood based, many of the larger CDCs operate in multiple adjacent neighborhoods. As a result, most of the low- and moderate-income areas of the city are within the jurisdiction of one of the core CDCs (see Lurcott & Downing, 1987, p. 460; PPND, 1990).[37] Two citywide organizations support Pittsburgh CDCs: the Working Group on Community Development, a coalition of community-based organizations that sets general policy strategy, and the Pittsburgh Community Reinvestment Group, a Community Reinvestment Act coalition that seeks to increase the level of investment made by financial institutions in CDC neighborhoods (see Jezierski, 1990; Metzger, 1992).

The PPND is a private organization acting as a financial/technical assistance intermediary between the CDCs and their primary funders: Pittsburgh's five major banks, local and national corporate foundations, and the city. Governed by a board of directors primarily composed of representatives from these entities, it commits between $2 million and $3 million annually in support of community-based economic development. Approximately half of this amount is used for CDC operating expenses; the other half, in the form of both grants and low-interest loans, flows to CDC development projects (Linsalata & Novak, 1992; PPND, 1990).

The city government, apart from contributing substantial sums of public dollars (primarily Community Development Block Grant funds) to the PPND, maintains additional funding streams to support CDC activities, money administered by the semiautonomous Urban Redevelopment Authority. It also provides technical assistance directly through the Department of City Planning and indirectly through its financial support of other community technical assistance organizations (Lurcott & Downing, 1987, pp. 462-463).

Viewed from a theoretical perspective, the overall Pittsburgh CDC effort parallels some of the key philosophical tenets of the community-based strategy outlined above. For example, according to the city's chief economic development officials, placing development capacity in the hands of community-based groups strengthens the autonomous ability of their neighborhoods to achieve economic viability, and hence "reduces the city's need to search out private investment for more than half the distressed neighborhoods in Pittsburgh." Moreover, the city benefits because these groups "are uncovering investment opportunities that the private and public sectors previously overlooked" (Lurcott & Downing, 1987, p. 464). Similarly, the Pittsburgh Partnership For Neighborhood Development's annual report more explicitly articulates the logic behind the claimed effectiveness of this "alternative" urban economic development strategy:

> We have seen approaches to urban renewal come and go, and we believe that *community-based development* [italics added] provides by far the best possible return on investment. The organized community has the greatest stake in the process. Its hands-on position is the best way to identify opportunities and needs, and to manage projects successfully. (PPND, 1990, p. 2)

And finally, a key official of the PPND and veteran of the Pittsburgh CDC movement sees the overall goal of the Pittsburgh CDC effort as a means to "empower" those now without power in the city.

Genesis. The story of how the city's extensive CDC development capacity came into being begins in the 1970s. During that period, neighborhood interests gained political incorporation into the governing coalition. This incorporation in turn worked to legitimize their demands regarding economic development (Sbragia, 1989, pp. 108-111). With active support from the city, several neighborhood-based organizations concerned with development issues emerged as significant actors by the late 1970s.

In 1981, the Allegheny Conference on Community Development—the key planning organization representing the local corporate community in the Pittsburgh regime—saw a need for the creation of several neighborhood development engines to buttress the overall economic health of the region. Subsequently, this organization embarked on a strategic plan to establish a network of CDCs in the city, mostly by upgrading the capacity of the existing neighborhood development organizations. To financially sustain these burgeoning CDCs, the Allegheny Conference, with the cooperation of the city government and local corporate foundations, arranged for the creation of the Pittsburgh Local Initiatives Support Corporation. Staffed out of the Allegheny Conference, this organization was a locally operating component of the Ford Foundation's national Local Initiatives Support Corporation program. In this arrangement, the national organization, a major conduit for CDC funding, matched money raised locally for the support of community-based development projects in Pittsburgh (Ahlbrandt, 1986, pp. 129-130; Lurcott & Downing, 1987, p. 462).

In 1983, building on its earlier efforts in the city, the Ford Foundation "selected Pittsburgh as a testing ground for local CDC-supporting intermediary organizations" (PPND, 1990, p. 7). The idea was to create a local organization that would supply a steady, reliable stream of operating money to CDCs, therefore creating more stable and professionally run community economic development bodies. Subsequent discussions among representatives of the Ford Foundation, local foundations, the Urban Redevelopment Authority, the Department of City Planning, and neighborhood development organizations led to the creation of the PPND. With public, private, and foundation dollars at its disposal, this intermediary organization channeled more than $1 million in operating funds to the city's five major CDCs over the next 2 years (Lurcott & Downing, 1987, p. 462).

In the late 1980s, this organization grew in three ways. First, its list of primary funders grew, as Pittsburgh's major banks joined the city and the foundations as key players. Second, the list of CDCs receiving operating support was expanded from five to ten. And third, the organization began also to serve as a conduit for project-related CDC funding.

CDC Development Activity. A brief profile of three of the more significant Pittsburgh CDCs provides an overview of the kinds of community-based economic development activities being undertaken in the city.

The North Side Civic Development Council (NSCDC) operates in the 14 neighborhoods of Pittsburgh's North Side, a section of the city across the Allegheny River from the central business district. The area is home to a little more than 40,000 people, most of whom have low or moderate incomes. Its economic base, a mix of residential, commercial, and industrial properties, suffered a bout of economic disinvestment in the 1960s, and the area has remained relatively depressed ever since (Peirce & Steinbach, 1987, p. 53; Osborne, 1988, p. 66).

Although this CDC engages in an array of economic development activities, its major efforts focus on the redevelopment and construction of commercial and industrial real estate. Over its first decade of operation as a CDC, it developed more than 400,000 square feet of commercial and industrial space. Projects include two buildings housing light industry, a small hotel, a restaurant, a brewery, and a commercial and office complex. More recently, the CDC assumed a role in a major waterfront development undertaken in its North Side service area. The organization secures funding and provides marketing for the many projects in this mixed-use development, and serves as a general partner for one of its buildings (NSCDC, 1991). The CDC has also been active in the realm of business development. Along with its general efforts to market the advantages of the area's location, it provides a variety of services to North Side enterprises, including direct financial and technical assistance. Through 1991, this CDC had worked with more than 80 small- and medium-sized companies. It also has provided the area with more than 100 units of new housing (NSCDC, 1991).

The Homewood-Brushton Revitalization and Development Corporation (HBRDC) CDC operates in the Homewood section of Pittsburgh, an area composed of three neighborhoods near the extreme eastern end of the city. Forty years ago, Homewood-Brushton was a middle- to upper-income, predominantly white area with a thriving commercial base and more than 30,000 residents. Today, it is an economically depressed, low-income area with fewer than 12,000 residents, most of whom are African American (HBRDC, n.d.; Peirce & Steinbach, 1987, p. 29).

This CDC has been very active in acquiring property along the area's main commercial strip, Homewood Avenue. The blighted condition of the area has allowed the CDC to purchase buildings and property relatively cheaply. By 1992, it controlled nearly 42,000 square feet of commercial space in this area (Linsalata & Novak, 1992). Through its rehabilitation and construction efforts, the organization has redeveloped much of this space. These efforts, along with its business assistance programs, have stimulated new business activity around Homewood Avenue, including three new retail franchises, a barbershop, a video production company, and a farmers' market. The CDC has also been heavily involved in residential real estate development in Homewood. Through both rehabilitation and new construction, it developed nearly 100 housing units. In addition, the organization publishes 20,000 copies of its weekly newspaper, and prints the newspapers of two other city neighborhoods. It also owns and operates a radio station.

Finally, the Oakland Planning and Development Corporation operates in the five neighborhoods of the Oakland section of Pittsburgh. Its population, numbering about 26,000, is a mix of incomes and races. Likewise, its relatively healthy economic base is a mix of institutional, commercial, and residential development. Oakland, home to the city's two major research and educational institutions—the University of Pittsburgh and Carnegie Mellon University—as well as to several hospitals, museums, and businesses, is the third-busiest area in the entire state of Pennsylvania, after the downtown sections of Philadelphia and Pittsburgh.

The Oakland CDC's development activities focus heavily on residential real estate development. In its first 10 years of operation, the CDC developed a total of 250 housing units, "a rate considered prodigious for a community development corporation" (PPND, 1990, p. 11). It also provides assistance services for neighborhood businesses and publishes a community newspaper.

This CDC has used its planning capacity as a means to increase its role in local development. With two major research universities, Oakland became a target area for significant development activity in the mid-1980s, as the city's governing coalition reordered its development efforts to focus on the creation and growth of advanced technology. In response, the CDC sought "to get a 'piece of the action' and thus to ensure that its moderate-income constituency would receive at least some benefits from the anticipated economic growth in Oakland" (Weiss & Metzger, 1987, p. 475). It commissioned a "technology impact analysis" to assess job creation potential, occupational structure, and locational requirements of those sectors considered most important to the advanced technology development effort in Pittsburgh and Oakland. With this planning analysis in hand, the organization gained "visibility and legitimacy" as a player in the local high-technology policy arena. As a result, OPDC was able to carve out a larger role for itself as a developer and to obtain additional benefits for area residents (Weiss & Metzger, 1987, pp. 472-476).[38]

Analysis of Impact

Despite some impressive accomplishments made by the city's network of CDCs, the overall significance of community-based economic development in Pittsburgh remains fairly marginal. The aggregate amount of economic activity controlled or influenced by community-based organizations makes up only a small percentage of the local economic base. And most economic development controlled by CDCs is inherently nonviable economically, and hence requires heavy subsidization. As we will see below, a number of powerful constraints inhibit the fuller development of community-based economic development in Pittsburgh. As a result, these efforts are unable to have the effects on the structural context of the urban regime necessary to make them vehicles of urban regime transformation.

In the analysis below, I attempt to isolate the conditions that limited the success of Pittsburgh's community-based economic development efforts. To clarify this discussion these conditions once again are classified as political, institutional, and economic.

Political Factors

The first and most straightforward political factor limiting the success of community-based economic development in Pittsburgh stems from the general imbalance of political power in the city's governing coalition. As alluded to above, local corporate interests, historically acting through the Allegheny Conference, have maintained a privileged position in this governing coalition (or as it is referred to in Pittsburgh, the governing "partnership"). In contrast, the city's neighborhood movement, while incorporated into the governing coalition by the early 1970s, clearly has remained a junior partner with only limited influence in key decision-making arenas (Jezierski, 1990). Moreover, Pittsburgh's African American community has fared even worse. Though the political interests of elements of this community are expressed through the neighborhood movement, the black community as a whole has been excluded from the governing coalition altogether (Sbragia, 1989). Because the composition of the city's governing coalition determines the content of its urban development policy (Stone, 1987), the relative powerlessness of those political interests with a fundamental stake in community-based economic development—the neighborhoods and the African American community—has diminished the overall scope and depth of these efforts.

The most visible aspect of this phenomenon is that the planning and programmatic activities of community-based economic developers have been relegated almost exclusively to neighborhoods. These interests, as Jezierski (1990, p. 239) notes, "have not gained enough public status to be included in citywide or regional development" (see also Sbragia, 1989, p. 110). For example, community-based development organizations were not consulted in the process that produced Strategy 21, a regionwide strategic planning document serving as the blueprint for Pittsburgh's Renaissance III. Moreover, this document failed to include any of their development projects.[39] Similarly, according to a top official at the city's quasi-public Urban Redevelopment Authority, in that agency's "thinking" about development efforts, a strict separation exists between "neighborhood development," which is the domain of the CDCs, and "development of the downtown and other business areas in the city."

Hence, community-based developers lack a role in citywide and regional development, and this reality severely handicaps community-based economic development efforts for at least two reasons. First, without this role, these developers cannot have an impact on the wider economic forces affecting their economic interests. They are, in short, left with little control over the direction and substance of the local development process. Yet without this control, previous development patterns are unlikely to be altered, and community-based economic development projects based in poorer neighborhoods will continue to be economically nonviable or, at best, marginal (cf. Duncan, 1986). Second, the range of investment opportunities available to community-based economic developers is small. Most important, community-based development actors have extremely limited access to the more lucrative development markets outside the narrow confines of economically depressed neighborhoods.

The constitution of the city's governing coalition is not the only political factor inhibiting community-based economic development efforts in Pittsburgh. Community-based economic development also is hampered because it has only limited support *within* the political groups that have a fundamental stake in it. The source of much of this opposition can be traced to the particular development agendas pursued by many Pittsburgh CDCs. For example, several CDC executive directors interviewed reported that their neighborhood economic development activities at times conflicted with the demands of residents and neighborhood activist organizations. Pittsburgh city planners Lurcott and Downing (1987) note, similarly, that Pittsburgh CDCs

sometimes "have found themselves proposing or supporting development proposals not supported by the neighborhoods at large" (p. 465).

To begin to understand *why* Pittsburgh CDCs often pursue the kind
of development agendas leading to this observed dissension, we must
recall that the impetus for the creation of the network of CDCs in
Pittsburgh came from top leaders in the local corporate community.
Seeking to protect their investments in the city, they saw a need for these
neighborhood development engines to strengthen the overall economic
health of the region. The subsequently created PPND—the key CDC
financial support entity—is very much an elite-level "public-private
partnership" effort in the tradition of other major development initiatives undertaken in post-War Pittsburgh, with banks, corporate foundations, and other corporate actors joining together with the local state
in common purpose (see Ahlbrandt, 1986).[40]

It is not surprising that the city's community-based economic development efforts and its traditional redevelopment efforts were initiated
and are controlled by, as one local observer remarked, "the same players." As pointed out in Chapter 2 (see also Stone, 1989), the current
structural arrangements in place in cities make tapping into the resource
base of the local business sector necessary to undertake complex policy
tasks such as community-based economic development. But the *effect*
of this control is that it privileges certain kinds of development strategies over others on the agendas of the CDCs.

In specific terms, these agendas reflect the development priorities of
the corporate elites on the PPND: real estate redevelopment and, to a
lesser extent, the start-up of technology-oriented companies. Yet as a
Pittsburgh CDC executive director explained, these economic development activities often "require a tremendous amount of staff, resources,
and effort . . . but have little to do with the people who live in the
neighborhoods." This, in turn, "creates tension between the CDCs and
other grassroots leaders and raises questions about who the constituency of the CDC really is." In fact, the constituency of the CDCs is
increasingly seen to be the members of the PPND, not the residents of
the neighborhoods in which they operate (Ferman, 1996, p. 102). One
CDC executive director admitted this relationship:

> We are as much an extension of that group of people [the partnership] as
> we are an extension of the neighborhoods . . . because of the way the

money flows. My budget is 60% to 70% funded by folks in one place—the CNG tower [which houses the offices of the partnership and the foundations] . . . pure and simple.

All of this works to weaken the base of popular support for community-based economic development in the city.

Institutional Factors

In the institutional realm, capacity was a salient impediment. Although, following the national trend, Pittsburgh's CDCs have built additional capacity over the years, they have not achieved the level of capacity necessary for a major expansion of their economic development role in the city. For example, as one of Pittsburgh's chief economic development officials commented, the capacity of the city's CDCs is often "stretched" when approaching more complex development endeavors. "One of the concerns that I have," he said, "is that they [the CDCs] are limited, and therefore they should not take on too much because they can't handle it." Likewise, as Metzger (1992, p. 94) explains, after the Pittsburgh Community Reinvestment Group, the city's Community Reinvestment Act coalition, secured a multi-million-dollar reinvestment agreement with Union National Bank, the possibility of similar agreements with other banks was hurt because it was believed such deals "would place pressure on CDCs to do projects that they did not have the capacity to initiate and complete."

The success of community-based economic development efforts in Pittsburgh also was limited because the ancillary community-based economic institutions of the Bruyn model—community land trusts, worker-owned companies, and community finance institutions—were not well developed in the city.[41] This restricted the aggregate amount of economic activity under the control of community-based organizations because, as discussed above, these institutions provide a community base to land, labor, and capital—the three key requirements of production.

Economic Factors

Economic conditions also have kept the community-based economic development efforts in Pittsburgh from having a greater impact. Most notably, after decades of disinvestment, the internal economic demand of most of the neighborhoods where the CDCs operate ranges from weak to very weak.[42] This in turn limits the economic viability of many community-based economic development efforts.

General urban economic trends such as commercial and residential suburbanization and economic restructuring have weakened the economic base of these areas. On the other hand, the Pittsburgh case strongly suggests a link between these economic factors and political factors.

A community-based development professional charged with the task of revitalizing a depressed area implicitly recognized this link when he dismissed the idea that these economic trends were the result of blind economic forces: "These predominantly minority and low-income neighborhoods did not get underdeveloped by accident; disinvestment did not occur automatically, as a freak of nature." More generally, the legacy of regressive redevelopment policies pursued by the city's governing coalition also damaged the economic environment in many CDC neighborhoods. One of these regressive decisions of particular significance involved the redevelopment scheme surrounding the construction of the Pittsburgh Civic Arena. To anchor the arena near the central business district, an entire low-income, primarily African American neighborhood adjacent to downtown, the Lower Hill, was demolished in the late 1950s. This renewal effort displaced 1,551 families and 413 businesses, without any planning for housing or resettlement (Lubove, 1969, p. 131). Not only did the project (and some surrounding expressways) cut off and spatially isolate the remaining African American community in the Upper Hill neighborhood from the downtown, the resulting displacement "made other black neighborhoods, such as Manchester and Homewood, more precarious because those displaced from the Hill crowded into these more stable black neighborhoods" (Jezierski, 1990, p. 230).

As alluded to above, the current political climate also damages the economic environment in CDC neighborhoods: The weak political position of neighborhood-based forces leaves community-based economic developers without a role in citywide and regional development, which in turn limits their ability to shape the local development process in ways favorable to their economic interests. A good illustration of this occurred when the Working Group on Community Development, the city's coalition of community-based organizations, attempted to halt the governing coalition's plans to expand the region's airport. Such an expansion, they understood, would divert investment and jobs away from the city (and their neighborhoods) and toward the suburbs. Yet this challenge proved unsuccessful, and the airport project went forward (see Ferman, 1992, p. 15).[43]

Moreover, some evidence exists that the specific development strategies pursued by the CDCs themselves present an economic barrier, as these strategies are not having the desired effect of economically strengthening the CDC neighborhoods. These strategies, as noted above, reflect the economic priorities of the CDC funders—that is, corporate actors in the Pittsburgh PPND—emphasizing first and foremost the redevelopment of the city's real estate infrastructure (with little or no equity participation by CDCs). Yet as one CDC official noted, some community-based developers feel strongly that, with such a heavy focus on physical redevelopment, they are pursuing the "wrong strategy." Another key CDC official added that the PPND "weighs real estate development more heavily because it provides a visible artifact," but that this can have an "insidious" effect on the neighborhoods and the community-based economic development process more generally.

Summary

The community-based economic development efforts in Pittsburgh faced numerous obstacles that limited their effectiveness. Most notably, the imbalance of power in the city's governing coalition had the effect of confining community-based development efforts to a narrow territorial focus in already disinvested neighborhoods. This limits the ability of community-based developers to influence areawide development patterns and to tap new, lucrative investment opportunities. In addition, the city's CDCs lack the necessary institutional capacity to expand their development efforts, and the ancillary community-based institutions of the Bruyn model are developed only rudimentarily. Finally, because the CDCs are dependent on the corporate-dominated PPND for their funding, they have limited control over their developmental agendas. This weakens their grassroots political support and, at the same time, hampers CDCs' efforts to strengthen the economies of their target areas.

Although the methodological limitations of case analysis prevent us from generalizing, a growing body of evidence suggests that the Pittsburgh experience parallels that of other cities, including Cleveland (Keating et al., 1989; Yin, 1994), Los Angeles (Twelvetrees, 1989), and Atlanta (Newman, 1993).[44] Thus, despite earlier evidence suggesting the community-based strategy's possible effectiveness as an economic development approach, the feasibility of implementing it on a scope necessary to alter the structural context of urban regimes remains seriously constrained. As in the case of entrepreneurial mercantilism, the obstacles seem to be quite formidable. The question of whether these impediments can be overcome—that is, the question of whether the impact of community-based economic development in cities can be broadened—will be revisited in the final chapter.

NOTES

1. The work of entrepreneurial mercantilism's key theorist, Jane Jacobs, on the dynamics of urban capitalism best illustrates this individualism (or "atomism") (see Jacobs, 1969; Polsky, 1988).

2. Along these lines, Bruyn (1987) sees a parallel between the "great transformation" from feudalism to capitalism chronicled by Karl Polanyi (1944/1957) and the spread of the new community economics movement: "New community-oriented firms are developing in small numbers that may alter the market system, just as the new commercial firms that were developing quietly in the sixteenth century altered feudal institutions" (p. 9).

3. These three traits of the community-based strategy—along with its hostility to both the capitalist market and the bureaucratic state—point to its affinity with the vision of urban anarchism as best expressed in the works of Murray Bookchin (see, for example, Bookchin, 1987, especially pp. 222-268; cf. Friedmann, 1982).

4. The basis for much of this activity is the model prescribed by the legendary Chicago-based community organizer Saul Alinsky (see Alinsky, 1969).

5. In his recounting of the experience of the Research and Development Division in the Chicago Department of Economic Development under Mayor Harold Washington (1983-1987), Giloth sketches one embryonic example of how a city can nurture and buttress community-based development institutions. The Washington administration, he concludes, "made a major contribution to the conceptualization and practice of collaboration between a municipal administration and the grass roots" (Giloth, 1991, p. 118).

6. Approximately 65 community land trusts have been formed in 20 states (Soifer, 1990, p. 240). They are found in a diverse array of U.S. cities, including Cincinnati (the Community Land Cooperative of Cincinnati); Dallas (Common Ground Community Economic Development); Boston (the Boston Community Land Fund); Atlanta (the South Atlanta Land Trust); and Burlington, Vermont (the Burlington Community Land Trust).

Many community land trusts have received considerable financial support from their local governments (see White & Matthei, 1987, pp. 44-47), including most notably the Burlington Trust, which was created with a $200,000 city appropriation (Soifer, 1990, p. 241).

7. This perspective owes much to the ideas of 19th century economist Henry George, as penned in his book *Progress and Poverty* (1879/1975). Also see Ross (1982, especially pp. 127-128).

8. Several cities have made efforts to assist the creation of worker-owned companies, including Burlington, Vermont (Schramm, 1987, p. 162); Pittsburgh (Beauregard, Lawless, & Deitrick, 1993, p. 425); St. Paul (Judd & Ready, 1986, p. 240); Chicago (Giloth, 1991, p. 109); New York; San Antonio; and Milwaukee (Rosen & Wilson, 1987, pp. 223-224).

9. The approximately 400 community development credit unions in the United States operate in cities including New York and Santa Cruz, California (Gunn & Gunn, 1991; Swack, 1987).

10. Community loan funds currently operate in cities including Boston; Hartford, Connecticut; and Atlanta (Gunn & Gunn, 1991; Swack, 1987). A more specific potential source of municipal financial commitments for these funds are repayments of Urban Development Action Grant loans made to private developers; many cities have dedicated that money to community economic development efforts (see Giloth, 1988, pp. 345-346; Peirce & Steinbach, 1987, p. 60).

11. According to Peirce and Steinbach (1987), "By 1987 community coalitions in more than 100 cities and towns had negotiated major CRA (Community Reinvestment Act) agreements with local banks, thus generating more capital for credit-starved neighborhoods" (p. 68). Public officials in cities have often played a strong supporting role in these efforts (see Squires, 1992b). Boston offers one example where Community Reinvestment Act challenges "helped to push two prominent banks to fund a Community Reinvestment Revolving Loan Fund, managed by a local community organization" (Schweke, 1983, p. 30). The rudiments of this approach to forging Community Reinvestment Act agreements are also in place in Chicago. There, the community groups bringing a challenge have "the charge of administrating parts of the program" (Wiewel & Weintraub, 1990, p. 171) set up to redress the problem, such as identifying and recommending appropriate loan recipients, and monitoring the recipient projects.

12. The most well-known example of a community development bank, the South Shore Bank in south Chicago, is the centerpiece of a "multipronged development instrument" (Taub, 1988, p. 5) known as the Shorebank Corporation. Shorebank is organized as a for-profit bank holding company; other subsidiaries focus on land development, minority venture capital, neighborhood housing, and job training (Giloth, 1988, p. 345). Community Capital, which opened in 1991 in Brooklyn, is another example of a community development bank. Although modeled on South Shore, its focus extends beyond a single neighborhood. It grants loans in a variety of low- and moderate-income communities throughout New York City (Roberts, 1992).

13. A particularly useful financial tool available to the latter allows governments to deposit public money in community development banks. These so-called linked deposit programs—in which public officials seek to achieve development objectives by placing public funds in selected institutions—are now common features of municipal finance (Rosen, 1988, p. 121).

14. For a similar view of community-based economic development, see Lane (1988), who lists worker co-ops, community development organizations, and credit unions as some of the essential institutions for community-based economic development. Also see

McArthur (1993, pp. 852-854), who lists credit unions, consumer co-ops, community development trusts, and community development corporations.

15. Thousands of consumer cooperatives exist in the United States today, many in cities. Common types of consumer goods and services supplied by co-ops include food, housing, and health care (see Hammond, 1987, pp. 100-101; Gunn & Gunn, 1991, pp. 99-103).

16. An excellent example of this phenomenon occurred in Baltimore. With some financial support provided by the city government, community-based activists sought to create a consumer cooperative that would sell auto insurance to city residents. These city residents pay nearly double the rates charged area suburbanites (see Chalkley, 1991).

17. The literature on CDCs is voluminous. For overviews, see Blakely (1989, chap. 10), Osborne (1988, pp. 303-305), Peirce and Steinbach (1987), Perry (1987), and Zdenek (1987).

18. CDCs to date have received financial and other support from a variety of sources, including governments at all levels, and corporate, religious, and philanthropic organizations. Most important for our purposes, many cities have become deeply involved in the development of CDCs. As Peirce and Steinbach (1987) note, "cities have become . . . potentially [the] most crucial . . . partner of CDCs. . . . One can find cities indifferent to their CDCs. But the trend is toward active cooperation and support" (p. 60).

19. As Meehan (1987) adds, Schumpeter—the economic theorist who more than any other placed the efforts of entrepreneurs at the center of the economic development process—also did not "personify [this function] in the form of heroic individuals, and . . . believed that this function could be, and often was, exercised cooperatively" (p. 135).

20. Kelly (1977), for example, cites the results of David C. McClelland's comprehensive analysis, *The Achieving Society* (1961). McClelland analyzed more than 600 studies from more than 30 countries and concluded, for instance, that "individuals with high achievement motivation are less likely to be motivated to action by monetary rewards than individuals low on achievement motivation," and that "the high achiever is *not* more motivated by individual than by group [i.e., collective] goals" (Kelly, 1977, p. 17).

21. Certain *means* by which a city may fund community-based development, such as "linked-development" policies, may be legally suspect on a variety of grounds (see Herrero, 1991, pp. 13-16). There is nothing inherent in this strategy itself, however, that needs to be examined as a potential legal barrier inhibiting its prospects for success.

22. This extension would include tapping markets external to poorer areas (both inside and outside the city) by "exporting" goods and services to wealthier communities as well as to the public sector (see Daniels et al., 1981, pp. 182-183; Fusfeld & Bates, 1984, pp. 234-235, 245-246).

23. For example, the comprehensive survey undertaken by Vidal (1992) of 130 CDCs (the key institution of the community-based strategy) revealed that only 23% served "severely distressed" neighborhoods (1992, p. 81).

24. One bank executive in Milwaukee said, "It's good business; you don't lose money doing CRA. It's the right thing to do." A banker in New York said that his bank's Community Reinvestment Act loan portfolio "looks better than the traditional real estate portfolio. You can do this kind of thing prudently." Similarly, in Philadelphia, bankers "are finding that making [Community Reinvestment Act] mortgages to inner-city borrowers promises to strengthen their bottom line in the long run, while not hurting it in the short term."

25. Woolworth, a major merchandising company, has, for example, correctly perceived the rationality of these economic opportunities. It has established 450 of its 1,000 variety stores in predominantly minority areas. And on average, the operating profit

margins of its inner-city stores are a percentage point higher than its stores in suburban areas (Alpert, 1991, p. 170). Similarly, another retail activity showing much promise is food distribution. Because of the lack of competition, as one marketing analyst concludes, "a well-run supermarket can make a mint in [poorer] neighborhoods" (Alpert, 1991, p. 170; see also Giloth, 1988, pp. 347-348).

26. Though critical of the community-based approach to economic development, Porter (1995) explains this dynamic:

> Advances in transportation and communication may have reduced the importance of location for some kinds of businesses. However, the increasing importance of regional clusters and of such concepts as just-in-time delivery, superior customer service, and close partnerships between customers and suppliers are making location more critical than ever before. (p. 58)

27. The National Congress for Community Economic Development (NCCED) surveyed what it called community-based development organizations: "Such groups go by a variety of names." Most are called CDCs, but some are called neighborhood development organizations or are Neighborhood Housing Services network members (NCCED, 1989, p. 3). Given the broader, generic definition of CDCs used in this study and elsewhere (see Peirce & Steinbach, 1987, p. 12; Zdenek, 1987, p. 115), for our purposes these organizations can be classified simply as "CDCs."

28. After an exhaustive review of a decade and a half of evaluation research on CDCs, Perry (1987) concluded, "All of the systematic research . . . can probably be summarized in one significant conclusion: this approach to local renewal is effective, but efficiency in the use of this strategy is threatened by inadequate attention to management skills" (p. 196).

29. Similarly, a major national study of 130 CDCs in 29 cities recently reported, "Community-based development is ripe for expansion. . . . All but one of the 29 cities visited have the potential to increase the role of CDCs in community economic development" (Vidal, 1992, p. 137).

30. ESOPs are by far the most common institutional form of worker ownership. By the mid-1980s, more than 7,000 companies had ESOPs, of which at least 10% had a majority of their stock owned by employees. Collectively these 700 or so companies employed more than 1 million people (Rosen & Wilson, 1987, p. 213).

31. In particular, whereas Quarrey's (1986) study matched five or more similar non-ESOP companies to each ESOP, the General Accounting Office (GAO) only used one comparison company. As Swaine (1993) notes, perceptively, "This increases the possibility of random elements distorting the GAO's findings. The non-ESOP matched firm might be both super-profitable and super-productive, for example" (p. 303).

32. More recently, these co-ops have faced the constraints afflicting the entire domestic plywood industry, challenging their future prosperity (see Gunn, 1992).

33. The market share of the Spanish economy captured by these companies increased from less than 1% in 1960 to more than 10% in 1976. Sales increased at an average annual rate of 8.5% from 1970 to 1979 (Dahl, 1985, pp. 123-124).

34. According to an informal survey, tenants of the Burlington Community Land Trust agreed by a 4-to-1 margin that "the organization is a better landlord than most others, primarily because it tends to be more people-oriented" (Soifer, 1990, p. 245).

35. A wealth of literature documents that these efforts are well established (see Jezierski, 1990, pp. 237-239; Linsalata & Novak, 1992; Lurcott & Downing, 1987; Osborne, 1988, pp. 66-67; Peirce & Steinbach, 1987; Sbragia, 1989; Weiss & Metzger, 1987). In Vidal's

(1992) research involving 130 CDCs, for example, eight Pittsburgh organizations were included in her sample. This was considerably more than the number studied in any other city similar to Pittsburgh in size (see Vidal, 1992, p. 26).

36. Except where otherwise noted, this analysis draws on a series of interviews conducted during the summer of 1991.

37. The service areas of these 10 CDCs have a combined population of approximately 140,000, nearly 40% of the city's total. This CDC service area population is about equally mixed between whites and African Americans. The 10 "core" organizations are the Bloomfield-Garfield Corporation, East Liberty Development Inc., Hill Community Development Corporation, HBRDC, NSCDC, Oakland Planning and Development Corporation, Breachmenders Inc., Garfield Jubilee Association, Manchester Citizens Corporation, and the South Side Local Development Corporation (PPND, 1990).

38. The "technology impact analysis" allowed the Oakland CDC to get additional public and private support to develop moderate-income housing. Perhaps the most significant benefit, however, came through the use of the analysis' projections of increased demand for new office space in the area. This served as the basis for the organization to become a developer (with a private firm) of a major mixed-use development project in Oakland that included an office building, garage, school, and hotel (PPND, 1990, p. 11; Weiss & Metzger, 1987, p. 476).

39. However, the Oakland Planning Development Corporation was later able to use its "technology impact analysis" to gain a "piece of the action" of Strategy 21 (see above). This CDC was in a unique position because it operates in the neighborhood where many key institutions critical to the advance technology strategy are located (see Weiss & Metzger, 1987).

40. Representatives of banks, and to a lesser extent, corporate foundations and other corporate actors dominate membership on the 20-person board of directors of the PPND. Almost three quarters of the board come from these institutions. There are nine representatives from the city's major banks, three from corporate foundations, two from other corporate associations, two from the public sector, three from academia, and one from a local hospital (PPND, 1990).

41. Various attempts to develop these ancillary institutions have enjoyed little success. For example, the Steel Valley Authority, a regional body formed in the mid-1980s to revitalize the area's declining steel industry, attempted but failed to open worker- or community-owned steel mills (Beauregard et al., 1992, pp. 424-425). Moreover, another well-publicized effort, the City-Pride Bakery, began as a worker- and community-owned venture but because of undercapitalization had to be sold to a private company in less than a year (Kowinski, 1993, p. 130).

42. One partial exception is the Oakland Planning and Development Corporation; with its many institutional facilities, economic demand in the Oakland neighborhood has been relatively strong. This in part accounts for its ability to produce housing units at what the PPND (1990, p. 11) calls a "prodigious" rate.

43. More generally, the governing coalition's recent pursuit of an economic development strategy oriented toward advanced technology also illustrates this point, as it, too, works against the economic interests of CDC neighborhoods and their residents (see Fitzgerald & Simmons, 1991, p. 516; Sbragia, 1990, p. 60).

44. Compare, for example, the conclusion of Keating et al. (1989) regarding Cleveland's CDCs: These development organizations "have received what has been deemed possible by [the city's] power structure" (p. 139). Or similarly, compare Marquez's (1993, p. 293) conclusion for his study of three large Mexican American CDCs: "Political restric-

tions, inadequate funding, and the criticism of neighborhood economic development ... have combined to assure that CDC activities will amount to little more than a symbolic gesture." Also consistent with many of the findings here is the survey of CDC efforts provided by Giloth (see 1988, pp. 348-349) and the overview provided by Covington (1989, pp. 183-184). The latter writes, "The lack of an autonomous resource base ... creates a potential source of tension between CDCs and the constituency they are committed to serve" (Covington, 1989, p. 183).

THE MUNICIPAL-ENTERPRISE STRATEGY

Municipal enterprise is the third strategy for alternative economic development.[1] Not unlike the previous two strategies, it also embodies the potential to lessen the structural dual dependencies at work in central cities (thus creating the preconditions for reconstituting urban regimes in ways lessening the degree of local political inequality). Cities have traditionally owned (and sometimes made profits from) enterprises such as airports, convention centers, sports stadia, hospitals, public utilities, and mass-transit systems (Garber, 1990, pp. 11-12). The central objective of this third strategy is to promote local economic development by expanding city ownership and profit-making activity to nontraditional areas of the local economy (cf. Clavel & Kleniewski, 1990, pp. 211, 228).

Although all three of the strategies are experimental in nature, the municipal-enterprise approach is by far the least developed. Because so few examples of it exist, this strategy offers considerably fewer data than the previous two strategies did for us to evaluate its prospects for effectiveness and feasibility. Hence, less is known about this approach's potential to be the vehicle for lessening the dual dependencies. Nevertheless, this chapter presents the evidence that is available, following an explication of the strategy's vision and substance.

EXPLORING VISION AND SUBSTANCE

Vision

The vision of the municipal-enterprise strategy springs, in part, from the program forwarded by democratic socialists earlier in this century for reforming the workings of the capitalist economy. The democratic socialists thought that a more rational, egalitarian, and democratically responsive economic system could be constructed by transferring ownership of selected enterprises to the public sector. Because state-owned enterprises would be driven by a concern for the public interest rather than by the pursuit of private returns, this reformed economy ostensibly would be obtainable in a society in which key industries were under public control. This program of "nationalization," however, generally failed to bring about the desired consequences (see Carnoy & Shearer, 1980, p. 38; Lindblom, 1977, p. 112; Nove, 1983, pp. 167-172). In contrast, the architects of the municipal-enterprise strategy seek to avoid the pitfalls of nationalization by envisioning the creation of public ownership on a *decentralized* level. In doing so they follow the prescient guidance of the eminent British social theorist R.H. Tawney: In *The Acquisitive Society*, Tawney argued decades ago that the "objection to public ownership, in so far as it is intelligent, is largely an objection to over-centralisation. But the remedy for over-centralisation is not the maintenance of functionless property in private hands, but the *decentralized ownership of public property.*"[2]

An even more crucial element of this strategy's vision comes from the notion of the "empowered city" as theorized by Harvard law professor Gerald Frug (1980, 1984, 1987, 1988) and others (see Elkin, 1987; Garber, 1989; 1990; Norman, 1989). As envisioned by Frug, expanding local public ownership—that is, the exercise of property rights by municipal corporations—becomes the means to alter the condition of "city powerlessness." This condition is an artifact of cities' tenuous status in legal doctrine, which is itself a product of the gradual historical erosion of city power through judicial action (see Frug, 1980; Garber, 1990; Hartog, 1983). By increasing the role of cities as property owners (and profit makers), Frug writes, the municipal corporation can supplant the modern business corporation as the locus of decentralized economic power in society (Frug, 1980, p. 1128). This institutional reconstruction, in turn, weakens the basic public/private distinction in classical liberal legal doctrine emerging in the early 19th century to "distinguish municipal corporations from private corporations, which enjoy property rights, and from governments, which have public power" (Garber, 1990, p. 4; see also Frug, 1980, pp. 1099-1101).

For Frug, the ultimate purpose of empowering cities through increased municipal property ownership is deeply political: "Cities," he writes, "have served—and might again serve—as vehicles to achieve purposes which have been frustrated in modern American life." If empowered, "they could respond to what Hannah Arendt [1962] has called the need for 'public freedom'—the ability to participate actively in the basic societal decisions that affect one's life" (Frug, 1980, p. 1068; cf. Elkin, 1987). The conceptual image here is the restoration of the ideal of the Greek polis: Free and equal citizens coming together in a genuine public sphere to settle the public's business through face-to-face discussion.

Finally, the vision of the municipal-enterprise strategy is also indebted in part to David Osborne's recent conceptualization of a "reinvented government" (see, for example, Osborne & Gaebler, 1992). Reinvented government, he writes, is the product of an "entrepreneurial spirit" that is "transforming the public sector." A key component of this reinvention is the development of what he calls "enterprising governments," which "earn rather than spend." As envisioned by Osborne, these kind of governments, based primarily at the local level, "turn . . . the profit motive to public use" by engaging in ownership and other economic activities that garner substantial public returns, as well as achieving other public goals (Osborne & Gaebler, 1992, pp. 198-216; cf. Clarke & Gaile, 1989; Lassar, 1990).

Substance

Cities pursuing the municipal-enterprise strategy undertake a variety of economic development activities. A city can pursue this develop-

TABLE 5.1   Paths for Municipal Enterprise

Direct Public Ownership (of economic enterprises)

Description: Ownership, in direct form, of economic enterprises
Benefits: Credit/investment sources, revenue sources, anchoring of business,
               less need for locational incentives
Examples: City banks, city insurance companies, sports teams, cable TV, hotels

Retained Public Ownership (of productive assets)

Description: Accumulating public property for lease
Benefits: Revenue source, greater public control, less capital mobility
Examples: Capital facilities, land

Public Equity Holdings (in private enterprises)

Description: Partial public ownership of private enterprises (via public-private
               partnerships or venture capital)
Benefits: More equal public-private partnerships, local business stimulation,
               high public returns
Examples: Malls, hotels, office buildings, high-tech firms

Other Nontraditional Public Enterprise Activities

Description: Short-term public ownership
Benefits: Public profits
Examples: Buying and selling land

---

ment approach at its most general level—establishing local public ownership of economic enterprises and other profit-making productive assets—via several specific paths. I describe and explain the essence of these paths below.

Direct Public Ownership (of economic enterprises)

The first and most straightforward way a city could pursue this alternative economic development approach is through direct municipal ownership of economic enterprises not traditionally owned by cities. Frug (1980, pp. 1150-1151; 1984, pp. 687-688), for example, stresses above all the importance of creating city banks and city insurance companies.[3] In doing so, he emphasizes that these institutions are significant sources of credit and investment capital: "Both city banks and insurance companies, as major lenders, could significantly affect the growth and nature of the city economy by changing the criteria for the selection of eligible borrowers" (Frug, 1980, p. 1150). Moreover, he adds: "If cities became major lenders to economic enterprise, they could influence decisions about plant closings, the . . . organization of work, the kinds of . . . products . . . manufactured, and the job opportunities afforded . . . [to] minorities" (Frug, 1984, p. 687). Finally, Frug points to the "exercise of power by [banks and insurance companies]" in the fiscal crises of New York and Cleveland as "indication of their current importance in cities' future" (Frug, 1980, p. 1151, note 399; see also Shefter, 1985; Swanstrom, 1985).[4]

Frug (1984, p. 689) also suggests city ownership of profit-making businesses. This activity could generate income (public profits), providing cities with an alternative source of revenue, as well as serve as the means to "create an alternative style of business organization." An even more important benefit for our purposes would be the "anchoring effect" that city ownership would have on potentially mobile businesses. This in turn leaves cities less vulnerable to disinvestment, and limits their need to dole out costly incentives to prevent business flight.[5]

Retained Public Ownership
(of productive assets)

A second way a city could employ this approach is by accumulating public property for the purpose of leasing it to private for-profit entrepreneurs—but with a provision for the retention of public ownership. This approach springs from the widely accepted proposition that "property is not a thing but a 'bundle of rights' " (Elkin, 1987, p. 178; Garber, 1990, p. 11), "some of which can be retained by the community for its own protection" (Lynd, 1989, p. 19).

Publicly owned property would include productive economic assets, such as capital facilities (buildings, plants, equipment, and so on) and, especially, land (Babcock, 1990, p. 12). Public land might be accumulated via the creation and operation of a "public land bank"—an urban institutional innovation commonly used in Europe and Canada but "vastly underutilized" in U.S. cities (Soifer, 1990, p. 239; see also Blakely, 1989, pp. 141-142; Cullingworth, 1987, pp. 186-172; Garber & Imbroscio, 1996, p. 614). The retained ownership of these productive assets provides the city not only profit-making opportunities (through leasing arrangements)[6] but also greater public control over how assets are used (Cullingworth, 1987, p. 171; Sagalyn, 1990, pp. 143, 151).[7] And insofar as local public ownership is extended to mobile economic resources, the city could reduce the economic pain caused by capital mobility.[8]

Public Equity Holdings (in private enterprises)

A third way cities can pursue the municipal-enterprise strategy is by gaining public equity holdings in what are essentially private enterprises (see Fainstein & Fainstein, 1995, p. 632). This affords the local public sector with partial ownership of these enterprises.

There are two principal means by which a city can gain such a holding. First, it can be an element of a "public-private partnership" designed to ease the implementation and perhaps ongoing operation of an urban development scheme. These partnerships and the development schemes they beget are now common features of the urban milieu (see Fosler & Berger, 1982; Frieden & Sagalyn, 1989). Yet because the public sector's role in these arrangements is usually quite limited, partnerships have been of an "unequal" nature historically (see Barnekov & Rich, 1989; Squires, 1989). In contrast, when cities receive an equity stake in a development project in return for their investment or other publicly provided benefits, this becomes a means of moving toward public-private partnerships "formed along more equitable and [for the public sector more] financially lucrative lines" (Cummings, Koebel, & Whitt, 1989, p. 202; see also Fainstein, 1990, p. 43). Profits (or "net cash flows") accrue to the public as well as the private sector.[9]

A second means by which a city achieves an equity position in a private enterprise is through the provision of venture capital. In return for this capital, cities receive equity in these companies.[10] This activity not only stimulates the development of new and growing local businesses by providing them with needed capital (see chapter 3); it also affords cities the opportunity to reap the high returns that private venture capitalists now claim (see Eisinger, 1991, p. 72).

Other Nontraditional Public Enterprise Activities

A final way that cities can pursue the municipal-enterprise strategy is by undertaking economic development activities that, unlike the above varieties of the concept, involve only short-term public ownership of productive assets. This broad category of activities includes most notably the buying and selling of land for a profit. [11]

ASSESSING PROSPECTS FOR EFFECTIVENESS

To begin the evaluation of this third strategy's potential to lessen the structural dual dependencies, I examine its prospects for effectiveness. As in the previous two chapters, in this exercise I attempt to marshal empirical evidence showing the strategy's promise for creating sustained economic development and vitality.

Overarching Issues

As noted in previous chapters, two important overarching issues are a strategy's relationship to orthodox economic theory and to established legal doctrine. Recall that as "alternative" approaches to urban economic development, the three strategies each challenge the prevailing economic orthodoxy—neoclassical theory. If this challenge proves to be unsuccessful—that is, if orthodox economic theory is indeed correct regarding these disputed matters—then this would limit significantly a strategy's likely effectiveness as a development strategy. Also recall that if the courts are likely to disallow key elements of a strategy, this repudiation similarly reduces its prospects for effectiveness.

Economic Theory

Of our three "alternative" approaches to urban economic development, the municipal-enterprise strategy challenges the prevailing economic orthodoxy most directly (cf. Walsh, 1978, pp. 14-16). This challenge centers on two key issues.

First, according to orthodox economic theory, public ownership and control of economic enterprises is inherently inefficient compared to private ownership and control. These inefficiencies *supposedly* exist for at least three interrelated reasons: (a) Directors of state-owned enterprises lack the incentives to manage their operations and resources wisely; (b) the public sector in general is overly bureaucratic, making managerial decision making for public enterprises slow and cumbersome; and (c) state ownership politicizes the economic sector to a high degree, which in turn leads to politically popular but economically unsound development decisions.

Cogent objections can be raised to counter the argument that public ownership is inherently inefficient. Bureaucratic decision-making structures plague the private as well as the public sector (Goodsell, 1985, p. 49). The more important question is whether the public sector can act in an entrepreneurial fashion, using its resources in ways heightening their productivity and efficiency. Evidence suggests this can in fact be the case, especially at the more decentralized levels of government where the smaller scale allows for the necessary flexibility and innovation (Osborne & Gaebler, 1992; cf. Clarke & Gaile, 1989; Eisinger, 1988). An entrepreneurial local government, in turn, can restructure managerial incentives—by, for example, linking compensation to enterprise performance—so as to encourage the productive and efficient use of publicly controlled economic resources (Osborne & Gaebler, 1992, pp. 138-165). Finally, the effect of politicization is also unclear; certainly in this era of fiscal retrenchment the local state has been able to cut popular services (and public jobs) such as police and fire protection, and public libraries.

Moreover, a closer look at the key theoretical argument demonstrating the relative inefficiencies of public enterprise reveals its basic weakness (Denning, n.d.). According to this line of economic analysis, which stems from attempts to develop a theory of the business firm, inefficiencies result from a dispersed ownership structure—whether in a public enterprise (held by citizens) or a private corporation (held by investors). Such a structure of ownership, it is theorized, often allows unproductive managerial practices to go unchecked.

In this setting, the private, investor-owned corporation has the advantage of allowing for the transfer of ownership rights in a (stock) market. This transfer, in turn, allows for a concentration of ownership (giving the relatively fewer owners better incentives to monitor managerial performance), while, at the same time, subjecting managers to market pressures (through fluctuating stock prices).

On the other hand, the public company is subject to a variety of accountability mechanisms the private company is not (such as independent performance audits, open-door requirements, open access to records, and statutory mandates). In addition, the public company often receives intense scrutiny by the press, public interest and consumer organizations, and elected officials. "Compared to their private counterparts," writes Denning, "public firms exist in a fishbowl." Therefore, he concludes, although the *incentives* to monitor managerial practices are greater in the private corporation, the *costs* of such monitoring are less in the public sector. "Which of the two factors is the more important for the productivity of the enterprise cannot be answered within the logic of . . . [this] economic method (Denning, n.d., p. 14).

The municipal-enterprise strategy's second challenge to orthodox economy theory cuts even deeper. Theoretical orthodoxy holds that— given the nature of a free market system—if the private sector fails to engage in an economic enterprise, no economic justification exists for the state to engage in such an enterprise (unless the market is "failing" in the conventional sense).[12] Otherwise, the entry of the public sector into an economic activity perforce must be economically irrational: In essence, the market has "judged" this activity to be unworthy of pursuit. The practical manifestation of this theoretical point is that the municipal-enterprise strategy raises the specter of "lemon socialism," where the local public sector owns enterprises not able to survive the rigors of the marketplace on their own.

This important issue can be addressed on two levels. First, as Osborne and Gaebler (1992, p. 216) ask, fittingly, "Where is it written that government should handle only lemons, while business gets all the profit centers?" Accordingly, the substance of the municipal-enterprise strategy outlined above shows there is no reason why the local public sector cannot pursue economic activities also pursued profitably by the local private sector (or in a "partnership" arrangement with the private sector). Hence, the concern about publicly owned "lemons" is an issue only for that subset of municipal enterprises shunned by private economic actors.

What about those shunned municipal enterprises? Is city ownership of such enterprises doomed by the logic of "lemon socialism?" Perhaps, but not necessarily: On this point, the tenets of orthodox economic theory can be challenged by developing the concept of what Alperovitz and Faux (1984, p. 147) and others call the "local public balance sheet." This alternative accounting system includes not only private costs and benefits of a given enterprise but "public" ones as well. Whereas a particular enterprise might appear to be unprofitable on a purely private balance sheet, once we factor in public (or "social") costs and benefits (e.g., the economic effects of unemployment on the city), the bottom line could look different (Luria & Russell, 1982, pp. 169-170).

Therefore, for at least some endeavors rejected by the private sector as unprofitable,[13] there might exist a strong economic rationale for the creation of a municipal enterprise.[14]

Moreover, there is one critical public economic cost to local communities often omitted by even accountants using a "public" balance sheet. This cost stems from the negative effects of interjurisdictional economic rivalry: As discussed above, local public ownership can save a community the (often considerable) expense of providing the economic inducements necessary to (a) prevent potentially mobile investment from fleeing to another place or (b) attract replacement capital to replenish lost investment. Such savings result because municipal ownership "anchors" enterprises in a particular spatial location.[15] Of course, insofar as any economic enterprise (private or municipal) requires a public subsidy to be viable, public money will need to be spent to retain or attract investment. In contrast, the savings to the public here result because mobile companies often extract overly generous inducements by playing competing communities against one another. In fact, many observers have conceptualized this context as an "arms race" (with its prisoners' dilemma game logic) in which competing cities "rationally" oversupply subsidies to mobile capital (see Levy, 1992; Peretz, 1986; Wolman, 1988).[16]

Legal Context

Relative to the entrepreneurial-mercantilist and community-based strategies for alternative urban economic development, this third approach operates in a potentially hostile legal context. Many scholars raise serious questions about the legality of the kinds of municipal-enterprise activities sketched above (see Babcock, 1990; Wegner, 1990). Frug, for example, argues that, according to existing legal doctrine, "municipalities may not engage in any 'business' activity unless it falls under the heading of a 'public utility' and is not for profit." (1980, p. 1065). This assessment, if indeed correct, leaves much of the municipal- enterprise strategy incompatible with existing law and subject to repudiation by the courts.

On the other hand, another prominent legal scholar, Professor Robert Ellickson (1982, pp. 1568-1571), convincingly challenges Frug's claim. Ellickson, who believes (and laments) that cities in fact can legally engage in business activities, launches a three-pronged attack. First, he argues that Frug grounds his analysis in legal doctrine prevailing at the turn of the century and now outdated: "During the twentieth century, state grants of power to cities have become more and more generous," writes Ellickson. "Dillon's Rule,[17] which required courts to construe strictly all state statutory delegations of power to cities, was widely accepted in 1910. Today it is a dead letter in most states" (cf. Schoettle, 1990, p. 67). In addition, during this century many states granted home-rule powers to cities, which "has not only given cities new powers, but has also created in them an (admittedly limited) right to resist state interference in their 'local' or 'municipal affairs'" (see also Babcock, 1990, p. 36).

"Second," he continues, "state courts have considerably altered their interpretation of the constitutional and statutory texts that they once invoked to limit city business activities." Most notably, the "public purpose" doctrines in state constitutions, which specify that cities may use their revenues only to carry out a narrow set of tasks, "continue to have some bite in most states, but less and less bite as the years pass" (see also Babcock, 1990, p. 36; Colton & Fisher, 1987, pp. 56-63; Schoettle, 1990, p. 64).[18] Today, notes Ellickson, "cities . . . rarely lose lawsuits that challenge their power to engage in business activities that deviate from the public utility strategy."

Finally, Ellickson makes his case by pointing to numerous examples of local governments engaging in business activities with legal impunity.

> Local governments currently develop housing complexes, retail stores, office buildings, sport stadiums, and redevelopment projects. They rent tools; own and operate distant vacation resorts; sell at retail products such as gasoline, liquor, light bulbs, and sportswear; and lend money to home buyers and business enterprises.[19]

Ellickson's eye-opening analysis helps us avoid the pitfall of uncategorically accepting the notion that the municipal-enterprise strategy lacks any legal foundation: Dillon's Rule has been weakened, home-rule grants have strengthened city power, the "public purpose" doctrine has been expanded, and finally, individual examples of apparently legal municipal-enterprise activities are plentiful. Nevertheless, federalism in the United States weds the legality of this strategy to the legal context of the various states. Therefore, generalizations concerning the issue are difficult to make with certainty (cf. Babcock, 1990, p. 36).[20]

One form of municipal enterprise severely restricted by state constitutions or statutes is the investment of public money in private enterprises in return for a public equity holding. After numerous municipalities went bankrupt investing in private railroads in the mid-19th century, many states made it illegal for cities to become involved in the financing of private enterprises (Babcock, 1990, pp. 34-35). These restrictions, however, often can be overcome if courts deem that such financing serves a "public purpose" (Mandelker et al., 1990, pp. 222-223). And as Ellickson (1982, p. 1569) noted, state courts have generally broadened their interpretation of what constitutes such a purpose (see also Schoettle, 1990, p. 64). Moreover, according to a recent legal analysis, any remaining legal problems inhibiting cities from achieving public equity holdings in private enterprises can be sidestepped through the creation of a local "public authority," formally independent from city government, to engage in such actions (Colton & Fisher, 1987, pp. 52-66).

A potentially serious legal hindrance to the practice of the municipal-enterprise strategy, in all its forms, stems from the tenets of federal antitrust legislation (the Clayton Act and the Sherman Act).[21] When a city engages in business activities and also uses its regulatory powers to influence the local development process, it exposes itself to charges of illegal anticompetitive behavior if these regulations negatively affect its competitors.[22] Though the Local Government Antitrust Act passed by Congress in 1984 exempts local governments from any liability for monetary damages, it still allows plaintiffs to be granted injunctive relief when courts find antitrust violations (Colton & Fisher, 1987, p. 67; Mandelker et al., 1990, p. 515). Such relief in turn could undermine a municipality's ability to play a dual role as a regulator/profit-making entrepreneur in the local economic arena. The general direction of the courts' thinking on this matter remains unclear at this juncture, however (Babcock, 1990, pp. 36-39). Future judicial action should determine the significance of this possible legal barrier for the development of the municipal-enterprise strategy.

Specific Issues

To assess further the prospects for this strategy's effectiveness, we now look at more specific research regarding the probable performance of municipal enterprises. Because, as noted above, the substance of this strategy remains so undeveloped, there is a dearth of past experience on which to base evaluative judgments. Therefore, the exercise below will remain limited and incomplete.

Interestingly, as Frug (1980, p. 1151) observes, among those hostile to the development of city ventures, some argue against proposed experiments on the grounds they are likely to perform poorly and fail; others on the grounds they are likely to perform well and succeed. Explaining, he writes:

> Those who argue that city enterprises will fail usually point to the municipal bankruptcies that followed city investments in railroads in the 19th century. . . . The argument that they will succeed has most frequently been expressed as part of an argument that if cities take over certain services, cities—being government bodies—will have an unfair competitive advantage over "private" interests who also provide those services (Frug, 1980, p. 1151, note 402).

Hence, perhaps because of our limited experience with this strategy, even those in agreement in their opposition to municipal enterprises differ drastically in their assessment of the likely performance of such ventures.

Much of the evaluative research that does exist on the public ownership of economic enterprises analyzes of the concept on a centralized level, for example, "nationalized industries." The economic performance of these enterprises has its defenders (see Carnoy & Shearer, 1980, chap. 2), but on balance, this performance can be confidently rated as relatively poor. Therefore, this national-level experience clearly suggests a caution. Yet the focus here, as noted above, is on decentralized versions of the concept.

A more useful body of experience instead comes from experiments with this strategy by states. Perhaps the most famous examples are two financial institutions established in the first quarter of this century—the publicly owned Bank of North Dakota and the Wisconsin State Life Fund, a publicly owned insurance firm. Both have compiled impressive track records. For example, although prohibited from making direct private and commercial loans, the Bank of North Dakota's "return on assets in the mid-seventies was double the rate of the best showing by any of the hundred largest private banks in the country" (Goodman, 1979, pp. 193-194). Similarly, the Wisconsin State Life Fund, because of low overhead and the absence of private profit, has been able to offer "premiums [that] are between 10 and 40% cheaper than comparable private policies" (Carnoy & Shearer, 1980, p. 69).

More recently, numerous states have engaged in public ownership by obtaining an equity position in private enterprises via the provision of venture capital to such companies. Although evaluative data on these programs are limited and incomplete (Eisinger, 1991, p. 71), some evidence attests to the worthiness of these efforts. According to one survey, state venture capital programs reported returns between 25% and 40% (a rate comparable to the high returns garnered by private venture capitalists), although it is unclear whether these figures are representative of all such programs (Eisinger, 1991, p. 72). A more in-depth investigation of one state's efforts also yielded a positive evaluative result. The author of this study, which examined a program in Connecticut, concluded, "State venture capital funds may indeed operate successfully and profitably." (Fisher, 1988, p. 175). Yet more recent and comprehensive research finds that the existing evaluations of these programs establish "neither that they are failures or successes as economic development tools" (Eisinger, 1993, p. 136).

On the local level, the thousands of municipally owned utilities stand as common examples of public ownership. In operation for decades, these utilities are generally acknowledged to perform quite well. They are efficiently run, and operate with management costs only about three quarters that of private companies. According to the Department of Energy and a recent industry study, these utilities deliver service at an average rate between 25% and 36% lower than their private counterparts (Goodman, 1979, pp. 195-196; Salpukas, 1995; see also Osborne & Gaebler, 1992, pp. 215-216).

A path-breaking book by David Osborne and Ted Gaebler (1992) gives high evaluative marks to other, more contemporary local-level examples of cities implementing elements of the municipal-enterprise strategy. Visalia, California, made money for 6 years owning a minor league baseball franchise—and then sold it for a profit (Osborne &

Gaebler, 1992, p. 216). Fairfield, California, made millions buying and selling developable land (Osborne & Gaebler, 1992, pp. 200-201). The municipally owned cable television system in San Bruno, California, charges a little more than half the rates of its private sector competitors *and* still returns 5% of its gross revenues to the city (Osborne & Gaebler, 1992, pp. 214-215). And the deal in which Santa Clara, California owned and eventually sold an amusement park will, according to estimates, leave the city with an income of at least $5.3 million per year in revenues after 15 years (Osborne & Gaebler, 1992, pp. 208-209).

Though Osborne & Gaebler's examples focus heavily on municipal enterprise in California, several studies that include a diverse array of settings also have reached positive evaluative conclusions. One of these studies found that Louisville's efforts to pursue equity ownership and public profit making in urban development partnerships with private actors were "a case in which the public sector extracted a share of the potential benefits in return for the amount of dollars it brought to the negotiation table" (Cummings et al., 1989, p. 220; see also Cummings, Koebel, & Whitt, 1988). Similarly, Hartford, Connecticut, realized many benefits in the 1970s by retaining municipal ownership of property leased to private developers (Clavel, 1986, pp. 30-36; see also Clavel & Kleniewski, 1990, pp. 210-211; Gunn & Gunn, 1991, pp. 140-141). And finally, New York City's retention of ownership of the land in a redevelopment project allowed "more than $1 billion . . . [in escalating rental] revenue . . . to be designated for low-income housing construction," which is an amount far exceeding the dollars typically allocated under traditional office-housing linkage programs (Fainstein, 1990, note 8).

Despite these findings, the available data on the municipal-enterprise strategy unquestionably remain too isolated and too inchoate to use as a basis for predicting its future prospects for effectiveness. More complete and useful data clearly await the wider implementation of this approach in contemporary cities.

UNDERSTANDING CONSTRAINTS ON FEASIBILITY

This observation brings us to our next question: What are the constraints inhibiting the fuller development of the municipal-enterprise strategy? As emphasized over and over in earlier chapters, if one of these alternative strategies is to serve as the vehicle lessening the intensity of the structural dual dependencies, it must be feasible to implement the strategy on a significant scale. Although the current underdevelopment of this strategy makes case analysis an unproductive means of addressing this question, considering the three general sources of conditions inhibiting the other two alternative approaches does point to some probable obstacles.

Political Factors

The extreme ideological hostility to the concept of public ownership persists as a salient political obstacle inhibiting the development of this strategy. Although all three strategies face this political barrier to some degree, the hostility to municipal enterprise is especially intense (see, for example, Conroy, 1990, pp. 103-117; Osborne & Gaebler, 1992, p. 215; Portz, 1990, pp. 114-121). Frug (1980; 1982) traces this ideological position to the dominance of liberal political thought in contemporary America with its strong conceptual dichotomy between the public and private, the so-called "public/private distinction." A "powerful restraint on city profitmaking business," he writes, "is the liberal conviction that profit-making is a private rather than public activity" (Frug, 1982, p. 1597). In short, many Americans—rightly or wrongly—remain philosophically opposed to the essence of this strategy.

Institutional Factors

Among the institutional barriers restricting the fuller development of this strategy, the limited institutional capacity of cities to engage in municipal enterprise stands out. In more direct examples of public ownership, the local state needs to possess the expertise, organizational strength, and competence either to manage a range of profit-making enterprises directly or oversee their management via contractual arrangements. Similarly, when municipal enterprise involves a partnership with the private sector (such as in cases involving public equity holdings in private enterprises), the local state committing public dollars must have the capacity to evaluate the merits and risks of often complex development deals (see Babcock, 1990, p. 39; Cummings et al., 1989). In both cases, the institutional capacity demanded would be significant.

Economic Factors

The fundamental economic dependence of cities would inhibit the development of this strategy in much the same way it inhibits the other two. This phenomenon is especially evident when cities attempt to pursue municipal enterprise through the reorientation of development partnerships with the private sector. As Cummings et al. (1988, p. 48) remark: "Being more aggressive in negotiating [municipal-enterprise] partnerships will, of course, put a city in a partially adversarial relationship with potential partners." Cities already in a weak market position relative to their competitors may not be able to afford exhibiting such a posture. This inability, in turn, constrains their ability to pursue municipal enterprise.

SUMMARY AND CONCLUSIONS

Although data on the municipal-enterprise approach are extremely limited, it seems to hold some potential as an effective urban development strategy, even in an era marked by a preoccupation with privatizing public assets and functions. Although in theoretical terms this strategy deviates substantially from the prevailing economic orthodoxy (neoclassical theory), how that might affect its likely effectiveness is unclear. Likewise, the strategy's development measures are unlikely to be largely disallowed, as some claim, by the courts. As for the more specific issues considered above, the brief discussion, though incomplete, also points in a positive direction.

As for the strategy's feasibility, we have identified a series of obstacles that are, once again, formidable. As was the case for the other two strategies, the ability of cities to implement this approach on a scope allowing it to alter the structural context of urban regimes remains in question. The possibilities for and likelihood of overcoming these impediments will be explored in the final chapter.

Notes

1. Sections of this chapter appear elsewhere in a much revised form (Imbroscio, 1995b).

2. Quoted in Schumacher (1975, pp. 252-253). More recently, Staughton Lynd made a similar case for the merits of decentralized public ownership (see 1987a, p. 957; 1987b, p. 15; see also Goodman, 1972, pp. 177-179).

3. Frug implies that newly created city banks and city insurance companies could be analogous to the public bank (the Bank of North Dakota) and the public insurance company (the Wisconsin State Life Fund) that already exist on the state level (Frug, 1980, p. 1150; see also Carnoy & Shearer, 1980, pp. 68-70; Goodman, 1979, pp. 193-196; Litvak & Daniels, 1979, p. 123). For another proposal on the state level for a publicly controlled and owned financial institution, see Zeitlin (1982).

4. To lower the exorbitant auto insurance rates paid by city residents, Baltimore's city leaders discussed the possibility of a city-run auto insurance company "chartered as an agency of the city government" (Frank, 1991; see also Chalkley, 1991). However, later plans for an alternative insurance agency in Baltimore called for the company to be set up as a consumer cooperative (see chap. 4). In 1981, Minot, North Dakota, attempted to establish a city-owned bank, but a ballot initiative was defeated at the polls (Ellickson, 1982, pp. 1572-1573).

5. Cities (and local governments more generally) have owned an array of profit-making businesses. Minor league baseball teams are one of the more common examples—Columbus (Johnson, 1991, p. 315); Toledo (Morris, 1991, pp. 89-90); Visalia, California (Osborne, 1985). Long Beach, California, owns and operates its own towing company, which makes a yearly profit. Santa Clara, California, owned an amusement park (Osborne & Gaebler, 1992, pp. 208-209). Although fitting the more traditional mode of local public ownership, many cities also own profitable cable television systems, including San Bruno, California (Osborne & Gaebler, 1992, pp. 214-215). St. Petersburg, Florida, owned a motel (Fisher, 1987).

6. "Participating lease" arrangements for the use of publicly owned property are now common. In these arrangements, a developer pays the public sector an annual base rent plus an amount pegged to project performance (e.g., private profits). As one economic development consultant explains, the principle at work is simple: "The more money the developer makes, the higher the rent" (quoted in Babcock, 1990, p. 14). Examples are found in Los Angeles, San Diego, New York, and Washington, DC (Babcock, 1990, p. 13; Fainstein, 1990, p. 45; Sagalyn, 1990, p. 143).

7. Clavel (1986, pp. 30-36) recounts how in the 1970s Hartford employed a real estate development strategy centered in part on the idea of the retention of city ownership and leasing arrangements. In a lecture titled "The City as a Real Estate Investor," City Council majority leader Nicholas Carbone, the architect of these efforts, said that the first lever was for the city to "try to own property and buildings. Such ownership would give the city control over land use and allow the city to realize the increasing value as land prices increased" (Clavel, 1986, p. 32; see also Bach et al., 1982, pp. 18-19).

8. An excellent example of this approach was devised by the Steel Valley Authority, a regional body formed by nine municipalities (including the city of Pittsburgh) to revitalize the steel industry in the Monongahela Valley of Southwestern Pennsylvania. According to its (ultimately unsuccessful) plan to reopen LTV Steel's electric furnaces in Pittsburgh, "the Authority would own the land and the buildings and would lease them to a for-profit entrepreneur for operation" (Lynd, 1989, p. 19). This retention of public ownership would protect the community against future private disinvestment.

9. Examples exist in Fairfield, California, 10% to 17% of net cash flow of a regional mall; Cincinnati, 17% of profits from a project that includes a hotel and a home office for a local bank; San Antonio, 17% of net cash flow from a hotel; and Louisville, 15% of net

cash flow from a project that includes a hotel and office building (see Babcock, 1990, p. 13; Cummings et al., 1989, pp. 208-214; Fisher, 1987; Osborne & Gaebler, 1992, pp. 200-202). Bowman's (1987a, p. 4) survey of 322 cities found that almost 19% obtained shared equity in projects in the past and more than 27% were likely to pursue this strategy in the future.

10. As Eisinger reports (1991, p. 67), states have been active as public "venture capitalists": Twenty-three states have implemented a total of 30 state venture capital programs. Locally, Kieschnick (1981, p. 381) describes how a public venture capital program in Buffalo purchased preferred stock in two advanced-technology companies. Fisher (1988, p. 176, note 4) also notes that other local venture capital funds exist in such cities as New Haven, Baltimore, and Providence.

11. Fairfield, California, for example, made a large profit by buying and selling developable land (Osborne & Gaebler, 1992, p. 200; see also Fisher, 1987). See Elkin (1990, p. 43), who suggests that development profits could belong to municipalities as a way of restructuring the local economic base. A case of a profit-sharing arrangement not involving public ownership is the law passed by the California legislature in 1992 allowing cities to draw on the profits of their ports. Los Angeles, for example, has proposed to fill its coffers with $44 million from its port's profits (Sims, 1992).

12. Browning and Browning (1983, pp. 23-50), Lindblom (1977, pp. 78-81), and Stokey and Zeckhauser (1978, pp. 297-308) summarize conventional treatments of market failure as presented in the welfare economics literature.

13. Some enterprises may not even be "unprofitable" in private hands but simply "not profitable enough" (Morris, 1982a, pp. 38-39; Osborne & Gaebler, 1992, p. 216). Many large corporations demand at least a 20% rate of return from their subsidiaries (see Lynd, 1987b, pp. 40-41). Alperovitz and Faux (1984, p. 149) tell the story of how a subsidiary of Sperry Rand in Herkimer, New York, that faced closure because it was not producing the 22% return required by Sperry, was purchased by a local community effort and earned a 17% return during its first full year of operation after the purchase.

14. Concerning public enterprise, Walsh (1978, p. 16) points out that "countertheories" to what she calls "classical economic theory" to "distinguish the functions of the private and public sectors . . . have not developed in the United States." Therefore, we have "no positive concept of public enterprise." In contrast, we might say that the local public balance sheet approach could represent such a countertheory.

15. As pointed out in the previous chapter, worker ownership can have a similar "anchoring effect" (see Olson, 1987).

16. One of the most interesting examples of footloose firms extracting generous concessions from competing cities occurs in the area of professional sports (see Euchner, 1992). Illustrating the advantage of local public ownership, Morris (1991, p. 90) explains how the Lucus County government purchased a minor league baseball team to prevent it from moving away.

17. "It is a general and undisputed proposition of law that a municipal corporation possesses, and can exercise, the following powers, and no others: First, those granted in *express words;* second, those *necessarily* or *fairly implied in,* or *incident to,* the powers expressly granted; third, those *essential* to the declared objects and purposes of the corporation—not simply convenient, but indispensable. Any fair, reasonable doubt concerning the existence of power is resolved by the courts against the corporation, and the power is denied" (as quoted in Mandelker et al., 1990, p. 91).

18. For example, the California Supreme Court (*City of Oakland v. Oakland Raiders,* 1982) held that city ownership of a professional sports team "satisfied a public purpose and was a logical extension of earlier cases holding that constructing a stadium passed

the public-purpose test" (Schoettle, 1990, p. 64). Moreover, in a case involving a city's use of its eminent domain power—the landmark Poletown case (*Poletown Neighborhood Council v. City of Detroit*, 1981)—the Michigan Supreme Court further extended the traditional notion of what city actions constitute a "public purpose" (see Hornack & Lynd, 1987, p. 121).

19. In addition, other scholars researching the question have reached conclusions similar to Ellickson's. For example, in her perceptive analysis of the issue, Garber (1990) also envisions judicial assent for Frug's ideas for municipal enterprises such as banks, insurance companies, and other profit-making businesses: "Legally," she writes, "Frug's suggestions for creating municipal resources [through public ownership] are not inherently problematic" (Garber, 1990, p. 11). However, one attempted *means* of acquiring municipal enterprises—"taking" them from private owners via eminent domain powers—has met considerable judicial opposition (Eisinger, 1988, pp. 321-328; Garber, 1990, p. 11; Weinberg, 1984).

20. For example, in North Dakota a proposal for a city-owned bank in Minot appeared to face "no serious legal obstacles" (Ellickson, 1981, p. 1572). In Michigan, however, a publicly directed effort to reindustrialize Detroit faced "stiff obstacles," including the fact that city-owned banks were unconstitutional (Hill, 1983, p. 119, note 33).

21. Following the doctrine articulated by the Supreme Court in *Parker v. Brown* (1943), states generally have been granted absolute immunity from antitrust liability. But in the *Lafayette* case (*City of Lafayette v. Louisiana Power & Light Co.*, 1978), the court ruled that the state's immunity does not apply to the actions of cities unless such actions are pursuant to a state policy. In the *Boulder* case (*Community Communications Co. v. City of Boulder*, 1982), the court held that even cities operating under home-rule powers are not immune. In the *Eau Claire* case (*Town of Hallie v. City of Eau Claire*, 1985), the court affirmed this stance but ruled that a city could obtain the state's exemption even if the city's actions were not expressly called for in a state statute or its legislative history (see Colton & Fisher, 1987, pp. 68-71; Elkin, 1987, pp. 175-176; Frug, 1988, pp. 306-307).

22. For example, a city might grant zoning approval for a shopping mall in which it has equity and then reject proposals for competing malls (Fisher, 1987; cf. Babcock, 1990, p. 38). Empirical analysis suggests, however, that cities are unlikely to restrict competition in this manner (Frieden, 1990, p. 50), perhaps because they stand to benefit from increased urban development through local jobs and taxation even without an equity stake in the additional projects.

PART III

Conclusions

6

TOWARD A RECONSTRUCTED CITY POLITICS?

The intellectual task of theorizing political reconstruction is always formidable and daunting. Identifying normative problems in the current empirical workings of political systems is one matter. Carefully specifying and designing theoretically grounded and politically practical reconstituted institutional arrangements to correct for these problems is quite another. Too often the former preoccupies political theorists, while the latter—the constructive enterprise—is, in a relative sense, neglected. This is a mistake. Despite the difficulties involved, theorists must begin to combine their normative criticism with a comprehensive analysis of institutional (re)design (cf. Elkin & Soltan, 1993).

This study is an attempt to theorize the political reconstruction of contemporary central city politics. It therefore faces the difficulties inherent in the task. After recapitulating the study's basic argument and findings, this final chapter works through some of the many remaining problems drawn out in the earlier analysis.

A RECAPITULATION

As shown in Chapter 1, contemporary central city politics in the United States is often animated by the existence of (a) a governing alliance between land-based business interests and local public officials that pursues (b) an urban agenda heavily oriented toward achieving economic growth in the city via so-called corporate-center/mainstream development strategies. Because this urban regime pattern is so pervasive, I labeled it the "dominant urban regime form." A discussion of the normative implications arising from the existence of this urban regime form revealed that its dominance greatly exacerbates the degree of liberal political inequality among the urban citizenry. This, in turn, works to undermine a key institutional buttress necessary for the larger liberal-democratic order to flourish. Finally, I pointed out that the rectification of this state of affairs is largely a matter of altering or "reconstituting" current urban regimes in specified ways.

To understand what is necessary to accomplish this reconstitution, we had to next inquire into the manner in which urban regimes form. The results of this inquiry (undertaken in Chapter 2) demonstrated that current urban regime formation can be traced to two broad structural features of the urban polity: the "external economic dependence" and "internal resource dependence" of city public officials. The former refers to the need for city officials to attract and retain sufficient amounts of external economic investment (i.e., either prospective investment capital or capital that is currently located in the city but is relatively mobile); the latter, to the need for those officials to access extrastate resources required for effective local governance. I argued that reconstituting urban regimes—and increasing (liberal) political equality in the city—can be achieved by lessening the intensity of the "dual dependencies."

Finally, Chapter 2 demonstrated that three alternative strategies for local economic development (entrepreneurial mercantilism, community-based economic development, and municipal enterprise) have the potential to reduce these dependencies. Yet because these strategies, when cities do use them, are still in the experimental stages, whether or how much they might be able to reduce the dual dependencies cannot be determined. I surmised that for a strategy to achieve this reduction (and associated goals), two criteria must be met: A strategy must be *effective* as a means for bringing about urban economic development and must be *feasible* to implement on a scope allowing it to have a significant impact on the structural context of urban regimes.

Chapters 3 through 5 analyzed these two questions empirically, after providing a detailed description of the vision and substance of each strategy. Regarding the first criterion, the possible effectiveness of the strategies was supported by a wealth of empirical evidence. Regarding the second, each strategy's feasibility was found to be beset with numerous formidable impediments.

I concluded that these strategies—although perhaps effective vehicles for urban economic development—cannot be feasibly implemented (on the necessary scale) in the contemporary urban context. Therefore, I begin the analysis in this final chapter by exploring whether the impediments constraining the fuller development of the strategies can be overcome.

OVERCOMING CONSTRAINTS ON FEASIBILITY

Of the three general kinds of impediments—political, institutional, and economic—constraining the feasibility of each strategy, political impediments appeared to be most significant, and so I begin with them.

Political Factors

Synthesizing the findings of early chapters, we find three interrelated political obstacles limiting seriously the future feasibility of the strategies.

Securing the Necessary Financial Resources. Pursuing entrepreneurial mercantilism, community-based economic development, or municipal enterprise (or some combination of the three) requires that revenue-strapped central cities be able to draw on financial resources currently beyond their political command.

This obstacle is not as substantial as it might first appear. If the local political will to do so can be mustered, much of the necessary money could be obtained by redirecting the substantial expenditures currently made by cities on corporate-center/mainstream economic development (cf. Nickel, 1995, p. 371). Moreover, as noted in earlier chapters, the strategies themselves potentially provide the means to tap innovative local sources of revenue, such as municipal pension funds and public profits from development.

Still, these sources are unlikely to provide sufficient money to implement the strategies on the broad scope required for the structural context of urban regimes to be altered significantly. Additional money must be secured from the two traditional resource pools for local economic development: the local corporate (land-based) business community and higher levels of government. But as we saw in the St. Paul case, the commitment of the local business community to "alternative" urban economic development is likely to remain limited. And as we saw in Pittsburgh, such commitment—when it does occur—may distort a strategy in ways hampering its fuller development.

That leaves the latter source—higher levels of government, particularly the federal government. The probability that higher-level governments will provide sufficient money and other resources to cities is also remote, however. As pointed out in Chapter 1, the economic transformation occurring because of the decline of the Fordist system of accumulation has weakened the central state (see Scott, 1988, p. 175) and left it less able to bring about significant social change. Likewise, Clarke and Gaile (1992, pp. 187-188) refer, fittingly, to the present era of urban governance as the "postfederal" period, and highlight the political barriers militating against the potential for a large-scale national (or state-level) response. This potential, they argue,

> is weaker than in the past, thanks to ideological interpretations of economic restructuring processes that trivialize the role of cities in the national economy and undercut the rationale for national urban policies. Neither the perceived crises nor the political constituency concerns that drove urban policy more than 2 decades ago are as salient now as international crisis and competition issues. In the absence of alternative interpretations of economic trends, national and state policy makers will be reluctant to seemingly jeopardize economic productivity goals by directing resources to communities.

"Thus," they conclude, "there is every sign that the chilly climate for cities will persist in the near future" (Clarke & Gaile, 1992, p. 188).

Nevertheless, in more marginal ways some reordering of federal (and state) fiscal priorities to favor cities is conceivable. Even if relatively modest, such a policy shift could lead to further expansion of the strategies. Given the paucity of resources likely to be involved, however, this shift would need to be designed in ways that deliberately supported the alternative development efforts.

An excellent example of this type of action was the Clinton administration's proposal (later withdrawn) to create a $382 million program that would lend money and expertise to community-based financial institutions. Why couldn't this approach be expanded? Analogous federal urban development programs could be designed to aid other alternative efforts (cf. Alperovitz, 1993). For example, just as the federal UDAG program (1978-1989) required that funded development efforts be joint public-private partnerships, future programs could maintain that emphasis but also require that joint efforts provide municipalities with a strong equity interest in development projects (cf. Elkin, 1990, p. 43).

Finally, a return to the local scene reveals yet another possibility for securing the necessary resources, one that taps into the "structure-altering" potential of the three strategies.

One potential effect of these strategies is, as I said in Chapter 2, to lessen the "internal resource dependency" of urban public officials. This reduction entails a restructuring of internal resources in cities by either strengthening the resources of the local state or altering the distribution of resources in local civil society to augment the resources held by noncorporate interests. So if a strategy actually did begin to change the structure of internal resources in this manner, this slow building of an alternative resource base could be used to further develop the strategy. Moreover, if this additional expansion of the strategy has the same effect on the structure of internal resources, that would create conditions allowing for even more expansion.

Garnering the Support of Local Leadership. A second political obstacle that impedes the feasibility of the strategies is their need for the enthusiastic support of *urban* political leadership. We observed this phenomenon in St. Paul, as Mayor Latimer's indifference to the Homegrown Economy Project and his failure to put his considerable leadership skills behind it caused the experiment to flounder. In Pittsburgh, a similar, if less dramatic, outcome occurred, as the political leadership exhibited a more subtle indifference to the city's community-based development efforts by ceding responsibility and control of these activities to local corporate interests.

My discussion of the need to secure financial resources points clearly to the salience of this obstacle—obtaining the money requires garnering local political support. For example, redirecting the expenditures cities now make on corporate-center/mainstream development can occur *only* if the local political will to do so can be mustered. Likewise, to be successful, any federal effort to assist the alternative strategies by providing supplemental resources would have to be met receptively by local public officials. Moreover, the "structure-altering" potential of the strategies also requires an enthusiastic embrace from local leaders, because if the strategies cannot be properly instituted at an initial stage, it is unlikely that the necessary dynamic can be unleashed.

We saw in earlier chapters, however, that a broad array of political disincentives currently works against the notion that urban public officials will champion the alternative strategies with the required zeal. Perhaps most important, the current distribution of power in local governing coalitions leaves relatively powerless those political interests that have a fundamental stake in alternative urban economic development.

In addition, unlike corporate-center/mainstream development strategies, the alternative strategies are less likely to entail the kind of development efforts—such as large, privately developed land-use projects —that, as Elkin (1987, pp. 37) points out, "generate a stream of benefits that can be used to build public support by those seeking to hold onto or achieve city office." The natural divisibility and relative abundance of material benefits flowing from corporate-center/mainstream development efforts furnishes the means to build and maintain effective electoral organizations. Such benefits (profit opportunities, contracts, jobs, etc.) can be readily targeted to specific private development interests and other individuals, easing the flow of campaign money and other electoral assets to current or prospective city elected officials (Elkin, 1987, p. 38). Forgoing corporate-center/mainstream development strategies in favor of those derived from alternative strategies could mean serious electoral difficulties for urban leaders.

In light of these and other political disincentives, the likelihood that the strategies will be embraced by local political leaders is quite small. On the other hand, some aspects of the urban setting could counteract the political liabilities currently plaguing these strategies.

For example, as documented in a wealth of urban political economy literature, corporate-center/mainstream urban development efforts largely have failed to improve the objective conditions of cities (Elkin, 1987; Levine, 1988; Stone et al., 1991). As the ostensible general economic benefits of these approaches fail to materialize, their continued political legitimacy may be called into question (Goetz, 1994, p. 103; Smith & Judd, 1984, p. 189; Squires, 1994, p. 97), neutralizing some of the electoral advantage stemming from their pursuit. This could open the door for alternative development strategies of the sort considered here.

Another situation conceivably counteracting the political liabilities of the alternative strategies arises from the political forces unleashed by the process of "post-Fordist" economic restructuring discussed in Chapter 1. Relating the insights of Aglietta's (1976) work, Clavel and Kleniewski (1990, p. 226) explain the phenomenon this way:

> The Fordist system of large plants with stable, homogeneous work forces laid the conditions for the development of the large industrial trade unions. While this gave workers increased power over their economic conditions, it tended to separate the workplace from the residential community, thus separating the locus of workers' economic activity and struggles from the locus of their political activity and struggles. But an implication of the change to post-Fordism, with increasing numbers of small firms and a less stable and more heterogeneous, nonunion work-force, is that the locus (sic) of economic and political struggles may become more entwined (see also Mayer, 1991; 1988; cf. Katznelson, 1981).

Where this occurs, momentum builds for local public policies that address the concerns of community residents for economic empowerment (cf. Fitzgerald & Simmons, 1991; Jonas, 1993, pp. 402-403). In this environment, the city's economic development agenda can become politicized, and the legitimacy of corporate-center/mainstream strategies may be challenged. As a consequence, the "political space" necessary for a reorientation of local economic development policies would be further expanded.

What the strategies studied here offer is a potentially viable alternative formula for local economic development to fill any political vacuum that might arise (cf. Nickel, 1995, pp. 371-372). As the work of Todd Swanstrom (1985), among others, teaches us, from a political standpoint the availability of workable alternative strategies is crucial. Swanstrom chronicles how urban populism in Cleveland suffered serious setbacks when Mayor Dennis Kucinich opposed the city's corporate-center/ mainstream economic development policies without advancing an alternative. By failing to implement an alternative program, Kucinich lost considerable political support because he failed to reassure voters that his administration was attempting to combat economic decline (Swanstrom, 1985, pp. 151-153).

Finally, underlying all of this is the fact that, as many cities have become increasingly poorer and minority-dominated, this has created a popular constituency for local development efforts that are designed to benefit city residents more directly (cf. Goldsmith & Blakely, 1992, pp. 188-189; Judd & Ready, 1986, pp. 244-245; Reed, 1988a). This, too, works to the political advantage of the alternative strategies because they are explicitly designed to distribute benefits in this manner.

Offsetting Ideological Biases. Ideological biases against the alternative urban economic development strategies present a special political problem, one more deeply rooted than the two already considered. All three strategies imply a strong role for the local state in urban development endeavors—a role inconsistent with the philosophical tradition of "privatism" that has historically guided such efforts (cf. Barnekov & Rich, 1989, p. 213; Warner, 1968). Likewise, as we saw in earlier chapters, the strategies present strong challenges to the neoclassical orthodoxy in economic thought—something thoroughly ingrained in Western, and especially, American, belief systems irrespective of its validity (see, example, Fallows, 1994; cf. Heilbroner, 1988).

Whether these biases can be overcome remains an open question. Evidence suggests, however, that they can be. This evidence is derived from various experiences with the municipal-enterprise strategy, where —with its emphasis on local public ownership of productive economic assets—ideological hostility is especially intense (see Chapter 5).

Recounting the efforts in Youngstown and Pittsburgh in the 1980s to employ local public ownership to prevent the permanent shutdown of steelmaking facilities, Staughton Lynd (1987a, p. 953) notes, "Steelworkers were motivated [to advocate this strategy], not by a belief in the intrinsic virtue of public enterprise, but by their desperate need for jobs. . . . It is doubtful," adds Lynd (1987a, p. 945), "whether middle

American workers, who only a few years ago showered rivets on anti-Vietnam War demonstrators while building the U.S. Steel headquarters in Pittsburgh, would have supported the idea of public ownership in any other circumstances." Hence, while recognizing the ideological barriers hampering the use of public ownership as a local development tool, Lynd's work teaches us that under certain circumstances—when there is a "desperate need for jobs," which is the case in most central cities—such barriers may be less formidable.

Moreover, events in Baltimore show that other exceptional circumstances can generate support for municipal enterprise among those normally opposed on ideological grounds. Forced to pay exorbitant auto insurance rates, nearly double those of area suburbanites, more than 300 city organizations, including many "middle class, mainstream neighborhood groups" have welcomed a proposal described as "at the same time radical and oddly pragmatic: a city-run auto insurance firm" (Chalkley, 1991, pp. 18-19). Similarly, since 1980, people tired of paying soaring electric bills to private utilities have created or actively pursued the creation of publicly owned municipal systems in over 50 localities (Salpukas, 1995).

More generally, ideological barriers to municipal enterprise are being brought down by the revenue pressures faced by cities in what Osborne and Gaebler (1992, p. 199) call "an age of fierce resistance to taxes." In this environment, local public officials look increasingly to the profits generated by municipal enterprises to fund services (Alperovitz, 1993; Fisher, 1987). Likewise, implementing municipal enterprise can be seen as part of an effort to run local government "more like a business" (cf. Cummings, Koebel, & Whitt, 1988, p. 36; Fisher, 1987, p. 33), as local government becomes engaged in for-profit activities with its eye on the bottom line.[1]

Hence, if the very acute ideological biases militating against the municipal-enterprise strategy can be offset, it is likely that the same can be true of the other two strategies, where such biases are not as extreme.

Institutional and Economic Factors

We now move from the political to the institutional and economic realms. Earlier chapters showed that the institutional and economic impediments constraining the feasibility of the alternative strategies for local economic development stemmed largely from political impediments. Nevertheless, economic and institutional impediments each had an independent effect on the strategies' feasibility. Hence, these impediments, too, must be overcome if the strategies are to be implemented (on the necessary scale) in the contemporary urban context.

The need for local economic development agencies and organizations to possess a significant degree of institutional capacity stands out among institutional impediments. Cities require sufficient institutional capacity to implement more fully the entrepreneurial-mercantilist and municipal-enterprise strategies or, in the case of the community-based strategy, to assist community-based actors in extending the range of their economic development ventures.

Yet good reason exists to believe that the necessary capacity—although untapped—is already largely in place: In their struggle to cope with the competitive conditions foisted on them over the past few decades, cities have developed an increasingly sophisticated skill to intervene in the workings of the local development process (Mayer, 1988). Both the range of activity and intensity of effort in economic development matters have expanded dramatically (see Bowman, 1987a, 1987b; Clarke, 1993; Preteceille, 1990). These have augmented local policy expertise, managerial and administrative competence, and organizational strength (cf. Clarke & Gaile, 1989, p. 581; Rich, 1992, p. 167). Thus, this impediment appears to be more perceived than real.

Among economic impediments, the fundamental economic dependence of cities in the current structural context stands out. Dependent on attracting and retaining sufficient amounts of external economic investment (i.e., either prospective investment capital or capital that is currently located in the city but is relatively mobile), local public officials have a limited ability to pursue alternative economic development strategies. The nature of this constraint, however, varies from city to city and must be determined empirically in each particular case (see Swanstrom, 1988). Hence, its precise effect is difficult to assess.

The possibilities for overcoming this economic dependence are strongly wedded to the "structure-altering" potential of the alternative strategies discussed above. That is, as a strategy is gradually put into effect, to the degree that it succeeds, the economic dependence of a city would be gradually lessened. This reduction creates conditions that, in turn, allow the strategy's further expansion. So with respect to this impediment, the feasibility of these strategies is linked to their ability to lessen the dual dependencies, a question to which we now turn.

LESSENING THE DUAL DEPENDENCIES: NOTES ON THE PROSPECTS

We have come full circle: We are confronting once again the question of whether the alternative strategies could lessen the dual dependencies of urban public officials. This issue, as we just saw above, is crucial for assessing the potential of the alternative strategies to overcome the existing barriers to their broader implementation. More important, it is the cornerstone of our entire constructive project: As I have argued, altering the structural context of urban regimes by lessening these dependencies is the means by which the desired regime reconstitution can occur.

Unfortunately, as pointed out in Chapter 2, the embryonic state of the strategies makes it impossible for this study to assess whether they in fact could be the vehicles for lessening those dependencies. Nevertheless, the study did provide insights relevant to this issue that can shed light on whether this outcome is more or less probable.

The External Economic Dependency

Each of the alternative strategies is, by design, a means to reduce the external economic dependency of local public officials in cities. By stimulating indigenous economic activity, anchoring (potentially mobile) capital, lessening uneven development, preventing the leakage of economic resources, augmenting public revenues, and so on, these strategies can contribute to that goal (see Chapter 2). Yet this contribution is a contingent matter: It depends largely on whether these strategies actually work as urban economic development strategies— whether they are effective at bringing about economic vitality in cities. For if it turns out that these strategies are, in a fundamental sense, economically irrational, their implementation is unlikely to bring about the desired reduction in external economic dependency.

So whether these strategies might lessen economic dependency depends on whether they might be effective as development strategies. That question was, of course, examined in this study, and a large body of evidence presented in Chapters 3 and 4 and, to a lesser extent, in Chapter 5, did point generally to the conclusion that these strategies would be effective approaches to economic development.

The Internal Resource Dependency

As demonstrated in Chapter 2, each strategy potentially restructures resources in ways lessening the internal resource dependency of urban public officials on a city's land-based business community. The municipal-enterprise approach strengthens the resources of the local state, while the other two strategies—entrepreneurial mercantilism and community-based economic development—alter the distribution of resources in local civil society. The former is straightforward in its path toward lessening this dependency of urban public officials, as it affects the local state directly, making its ability to bring about the desired structural change relatively less problematic. The latter two, in contrast, affect the local state only indirectly by altering local societal dynamics. This characteristic makes their ability to lessen the internal resource dependency of urban public officials less certain.

Entrepreneurial mercantilism potentially accomplishes this alteration of local societal dynamics by promoting a diversification of the local economy through the proliferation of numerous small businesses. This diversification, in turn, both fragments the concentration of resources currently held by the city's corporate-oriented, land-based business community and creates the foundation for an urban regime rooted in what Stone (1989, p. 228) calls a "resource-rich but noncorporate middle class," and parallels Unger's (1987) notion of "petty bourgeois radicalism" (see Chapter 2).

Yet the examination of "small-business development" conducted earlier showed that lessening the internal resource dependence via this strategy confronts a major problem: Local small businesses often rely on larger businesses (Preteceille, 1990, p. 43); more to the point, many small businesses, as DeLeon (1992a, p. 157) notes, were founded to serve large corporations—and cannot survive without them. Therefore, in *political* terms, building a local economy of small businesses cannot possibly fragment the political resource base of a city's corporate community because—without an array of vibrant local corporations—a local economy constituted as such cannot exist.

The reasoning behind this argument can be challenged, however. Reacting to similar thinking, Case (1989, p. 44), for example, writes that this notion "makes sense only if we assume it's always big companies that 'place the orders' and that small companies can only respond to such markets. It allows no room for small companies buying from one another or for small companies creating new markets through innovation." Such thinking, he continues, stems from the distinction between basic and nonbasic industry—where large, basic industries export their goods and services outside the local economy and, hence, create opportunities for small, nonbasic companies. "But these days," Case concludes, "small companies as well as large ones fall into that [exporting] category." Moreover, as Eisinger (1988, chap. 12) demonstrates, state and local governments increasingly are finding that, with their help, local small businesses can even tap markets in foreign countries.

The community-based strategy, as we also saw in Chapter 2, potentially alters local societal dynamics in ways lessening the internal resource dependency of urban public officials by building a significant nonbusiness resource base in the city.

This base expands as the community-based, neighborhood-oriented sector procures the fruits of local economic development efforts. The existence of this community-based, neighborhood-oriented resource base reduces the dependence of local public officials on attracting the resources of the land-based business community, because it offers an alternative concentration of resources.

As the Pittsburgh experience with this strategy showed, however, a potentially major problem with this way of lessening the internal resource dependency is retaining the "alternative" character of these community-based resources. Consistent with the case findings presented above (see Chapter 4), many analysts of community-based development claim that once community-based activists take on economic development activities, they invariably become isolated from their grass-roots constituencies as they are co-opted into the city's "establishment" (cf. Covington, 1989, pp. 181-184; Cunningham, 1983, p. 264; Yin, 1994). At the heart of that establishment is, of course, the local land-based business community. So the nature of the control of the ostensibly alternative resource base that this development potentially creates might not be extremely different from what now exists. As a result, the resource dependence of local public officials on the local land-based business sector would remain intact.

On the other hand, while the threat of co-optation is indeed substantial, there seems to be no real alternative for community-based political power. So-called "protest" (or "advocacy") strategies undertaken by community-based actors, the usual alternative to engaging in economic development, though useful, have definite limits. In fact, to invoke a phrase familiar to students of urban politics, we might say that—for the purposes of regime reconstitution—"protest is not enough": Only by building a resource base will community-based actors gain a more central position in the city's governing coalition. Moreover, only through access to capital accumulation can the necessary base of resources be built (cf. Alperovitz, 1991).[2]

COMBINING STRATEGIES IN PRACTICE

For analytical purposes, this study considered the three alternative strategies as separate and independent routes for achieving urban economic development. This differentiation allowed us to better understand the essence of each approach, but in practice cities charting an alternative course in local economic development can (and are likely to) pursue these strategies in some combination.[3]

How might they be fruitfully combined in practice? The most obvious link would be to pursue entrepreneurial-mercantilist development goals—stimulating small-business development, increasing import substitution, tapping innovative local finance sources, and so on—in ways that nurture and buttress the alternative means of capital ownership and control embodied in the community-based and municipal-enterprise strategies. For example, innovative local finance sources would be used not just for any form of urban economic development, but for those forms that are community-based or municipally owned. Similarly, the stimulation of small-business development might focus on those enterprises wedded to community development corporations, or those that are partially or wholly owned by the city.[4] In short, the goals of entrepreneurial mercantilism would define the general economic development *strategy* of the city, and that strategy would be oriented in ways supporting the growth of the city's community-based economic institutions and municipal enterprises.

CONCLUSION: NORMATIVE MATTERS, AGAIN

This work has strived to be, first and foremost, a work in political theory —albeit one that is empirically based and focused heavily on constructive matters. We conclude, therefore, with a return to the normative issues from which it began.

In the introductory chapters, I argued that lessening the dual dependencies of local public officials creates the structural context necessary for the reconstitution of current urban regime forms. This reconstitution involves altering both the *who* and the *what* of central city politics: who exercises political power in the local decision-making process and what those with power attempt to accomplish. Specifically, it involves (a) making the governing coalitions of central cities more inclusive (see Stone, 1989, p. 243) and (b) reorienting the local policy agenda so that its main task is to expand economic opportunities for the lower class in cities (see Stone et al., 1991, pp. 232-236).

In turn, urban regime forms constituted as such were shown to be more consistent with the normatively desired condition of (liberal) political equality among the urban citizenry, where—following the work of Walzer (1983, 1984)—the success (or failure) of citizens in nonpolitical institutional spheres does not convert into success (or failure) in the political sphere (or vice versa). This consistency emerges because the reconstitution of urban regimes in the manner discussed above mitigates the two problems that current urban regime forms pose for the achievement of (liberal) political equality in central cities: privileged voices and economic inequality. Increasing the inclusiveness of a city's governing coalition conceivably corrects for the former because the influence in central-city politics wielded by land-based business interests is placed on a more equal footing with that of other political actors in the city. Reorienting the city's policy agenda so that its main task is to expand economic opportunities for the lower class conceivably corrects for the latter because this reorientation mitigates some of the worst aspects of economic, and hence political, deprivation among the poorest elements of urban citizenry (the so-called "urban under-class").

Achieving this political equality in central cities is a means of realizing our ultimate normative aim: Buttressing the foundation of one desirable political way of life, namely liberal democracy in the American context (cf. Elkin, 1987). This is because a liberal-democratic political order requires a decentralization of state power; yet systematic bias plagues the current state of political life in some of the local jurisdictions where power can be decentralized—that is, central cities. As a result, a devolution of state power in this context fails to foster the values of the larger political order. In short, decentralized state power must be properly constituted, and the reform of urban politics to achieve (liberal) political equality in cities accomplishes this aim.

Of course, any of the causal connections of this argument—though logically derived and empirically grounded—might not hold in practice. Hence, these normative aspirations would be left unfulfilled. I discuss and address three of the most salient problems with the argument below.

First, some urban analysts might question whether the "inclusive" governing coalitions envisioned by the urban regime of (liberal) political equality could be the basis for *effective* local governance. With a much greater variety of political interests in the city exercising influence—that is, with the more complete democratization of local mechanisms for popular control—centrifugal forces may undermine the effectiveness of political authority in the city (cf. Savitch & Thomas, 1991, pp. 244-250). Without this effectiveness, governing coalitions are unlikely to be stable and, at a minimum, will have great difficulty accomplishing complex policy tasks.

This perspective—though relevant—may be flawed. Most notably, it fails to grasp what might be called the Lindblomian point—that is, that there can be a basic "intelligence of democracy" (see Lindblom, 1965). Governing effectiveness—or the capacity for "social problem solving" (Elkin, 1987)—need not be inhibited by greater democratization; in fact, it can be enhanced by it (cf. Vogel & Swanson, 1993). Explaining this dynamic, Elkin (1987, p. 95) writes, "if the operation of popular control is systematically biased, problem solving is also likely to be ineffective, simply because some desirable alternatives will go unexplored. This is, of course, what is happening in cities." (see also Stone, 1989, pp. 211-212). Hence, a strong basis exists for believing that inclusive governing coalitions will not lack the necessary effectiveness.

A second possible problem is that even if it is agreed that inclusive governing coalitions generally can be the basis for effective urban governance, urban analysts may nevertheless raise questions regarding the other element of "reconstituted" urban regimes—the reorientation of the urban policy agenda so that its main task is to expand opportunities for the lower class. Here the issue involves whether such a reorientation can bring about what is for normative purposes required: a significant reduction in the degree of economic deprivation experienced by the urban underclass. Some may say that addressing such a deep-rooted and severe social problem requires a massive national-level policy response. Yet such a response, as noted above, is unlikely to be forthcoming.

This objection is a legitimate concern for anyone seeking to establish political equality in cities. It may, therefore, seem improbable that the normative aspirations of this study can be achieved. On the other hand, such a pessimistic view derives from a static—and hence inadequate—conception of the current status quo; in particular, it assumes that the ability of cities to bring about social change is unalterable (cf. Goldsmith & Blakely, 1992, pp. 179-182). Yet if the alternative strategies in fact can be successfully implemented, the restructuring of the urban political economy conceivably brought about by these strategies *itself* provides urban governing coalitions with some of the additional resources (and "economic space") needed to address the underclass problem.

Thus, it appears that until the necessary reconstitution of urban regimes occurs—until we have some relevant empirical experience by which to evaluate this objection—the matter perforce will remain unsettled. Certainly, however, the above pessimism is not completely unwarranted: Under any set of conditions, the task of significantly improving the plight of the urban underclass in the absence of massive federal resources is, to say the least, awesome. On the other hand, without changes in the current workings of central-city politics, federal efforts to address the underclass problem are likely to fail, given that current urban regimes are themselves a relentless source of material inequality (Elkin, 1987, pp. 99-101). Finally, we should remember that even if the problem cannot be solved, to the extent that it is ameliorated we can begin to move closer to realizing our normative goals.

The mechanisms used to bring about a lessening of the "dual dependencies" of local public officials—the three alternative strategies for local economic development—*themselves* present a third threat to our normative aspirations. Although these mechanisms may be successful in creating conditions that allow for both the development of inclusive local governing coalitions and the decrease in severe economic deprivation, the resulting urban regimes still may not be constituted in the ways necessary to buttress the foundation of liberal democracy. The democratization of decentralized public authority certainly would be achieved, correcting for the systemic bias of current urban regimes. But urban regimes reconstituted via the alternative strategies also could prove to be antagonistic to the individual rights and liberties so central to liberalism (cf. Elkin, 1987, p. 183).

The thrust of the entrepreneurial-mercantilist strategy to wall off or isolate the local community from exogenous forces represents one such threat. Historically, closed communities often have not been tolerant communities (cf. Dahl, 1967, p. 961; Young, 1990, pp. 234-236). Likewise, the enhancement of the power of the local state apparatus wrought by the municipal-enterprise strategy is a potential threat. A powerful state, even in a decentralized context, possesses the wherewithal to run roughshod over individual rights.[5] Moreover, elements of both of these dangers seem to be intrinsic to the community-based strategy, as its collectivism suggests both the isolation tendencies of entrepreneurial mercantilism and the concentration of power associated with the municipal-enterprise strategy.

Finally, if the dynamics described in this study began to unfold empirically, the messy world of political practice would no doubt present additional normative difficulties. Once again, however, the analysis of these problems can come only with relevant experience. If such experience in fact does begin to materialize, we can begin to better understand whether the normative aspirations expressed here can be realized. If these aspirations can indeed be realized, the contemporary American central city and the larger liberal-democratic order of which it is part will be reconstructed in a manner more worthy of its citizens.

NOTES

1. Viewing the strategy from this "businesslike" perspective, municipal enterprise has been strongly endorsed by ideologically mainstream thinkers such as David Osborne (see Osborne, 1985). Osborne's ideas concerning the "reinvention of government" made him a darling of the centrist Democratic Leadership Council (a national political group devoted to moving the Democratic party in a more ideologically conservative direction). "In reality," he and his co-author write, "there are several good reasons why government [by engaging in public enterprise] *should* sometimes compete with the private sector" (Osborne & Gaebler, 1992, pp. 216; see also Osborne, 1985).

2. The transfer of massive amounts of revenue and other resources from higher-level governments to local community-based organizations of course also would allow the necessary base of resources to be built. But the prospect of such a transfer taking place is highly improbable (see analysis above).

3. For example, the St. Paul entrepreneurial mercantilist effort to build a "homegrown economy" included plans for some community-based neighborhood enterprises and worker cooperatives, as well as some public equity holdings (see Judd & Ready, 1986, p. 240; Office of the Mayor, 1983). Likewise, the "alternative" local economic development plan pursued by the administration of Harold Washington in Chicago included elements of the entrepreneurial-mercantilist strategy and the community-based strategy (see Giloth, 1991; Judd & Ready, 1986, pp. 230-233).

4. Along these lines, Polsky (1988, p. 14) sketches a plan for blending what I have labeled entrepreneurial mercantilism and community-based economic development, suggesting a marriage between Jane Jacobs's inspired entrepreneurship and "a community-directed management structure." Likewise, Brown (1993, pp. 219-220) discusses alternative economic development efforts in the United Kingdom involving entrepreneurial mercantilism's emphasis on reducing imports and increasing multipliers, coupled with a system of ownership and control fitting the community-based model.

5. This threat is evident even if its authority is thoroughly democratized, as the "results of democratic processes, like all others, may be tyrannical" (Gutmann, 1983, p. 25).

REFERENCES

Aglietta, M. (1976). *A theory of capitalist regulation: The U.S. experience.* London: New Left Books.

Ahlbrandt, R. S., Jr. (1986, November). Public-private partnerships for neighborhood renewal. *Annals of American Academy of Political and Social Science,* pp. 120-134.

Ahlbrandt, R. S., Jr., & DeAngelis, J. P. (1987). Local options for economic development in a maturing industrial region. *Economic Development Quarterly, 1,* 41-51.

Ahlbrandt, R. S., Jr., DeAngelis, J. P., & Weaver, C. (1987). Public-private institutions and advanced technology development in Southwestern Pennsylvania. *Journal of the American Planning Association, 53,* 449-458.

Albelda, R., Gunn, C., & Waller, W. (Eds.). (1987). *Alternatives to economic orthodoxy.* Armonk, NY: M. E. Sharpe.

Alex-Assensoh, Y. (1995). Myths about race and the underclass. *Urban Affairs Review, 31,* 3-19.

Alinsky, S. (1969). *Reveille for radicals.* New York: Vintage.

Allen, D. N. (1989). Book review of David Birch's *Job creation in America. Journal of Regional Science, 29,* 297-298.

Alperovitz, G. (1990, July). Building a living democracy. *Sojourners,* pp. 11-23.

Alperovitz, G. (1991, May 31). [Photocopy of memorandum to Marcus Raskin.] National Center for Economic Alternatives, Washington, DC.

Alperovitz, G. (1993, Winter). A proposal . . . for a second Clinton beginning. *Newsletter of Pegs, 3,* 4, 15.

Alperovitz, G., & Faux, J. (1984). *Rebuilding America.* New York: Pantheon.

Alpert, M. (1991, July 29). The ghetto's hidden wealth. *Fortune,* pp. 167-174.

Alt, J. E., & Chrystal, K. A. (1983). *Political economics.* Berkeley: University of California Press.

Arendt, H. (1962). *On revolution*. New York: Viking.

Armington, C. (1983). *Further examination of sources of recent employment growth analysis of USEEM data for 1976 and 1980*. Washington, DC: Brookings Institution.

Armington, C., & Odle, M. (1982). Small business—How many jobs? *Brookings Review, 1,* 14-17.

Arnstein, S. (1969, July). A ladder of citizen participation. *AIP Journal,* pp. 216-224.

Babcock, R. F. (1990). The city as entrepreneur: Fiscal wisdom or regulatory folly? In T. Lassar (Ed.), *City deal making* (pp. 11-43). Washington, DC: Urban Land Institute.

Bach, E., Carbone, N. R., & Clavel, P. (1982, Winter). Running the city for the people. *Social Policy, 12,* 15-23.

Barber, B. R. (1984). *Strong democracy: Participatory politics of a new age*. Berkeley: University of California Press.

Barnekov, T., & Rich, D. (1989). Privatism and the limits of local economic development policy. *Urban Affairs Quarterly, 25,* 212-238.

Beauregard, R. A., Lawless, P., & Deitrick, S. (1992). Collaborative strategies for reindustrialization: Sheffield and Pittsburgh. *Economic Development Quarterly, 6,* 418-430.

Bechtel, D. (1989, April). Collective trust grows in the Lower East Side. *City Limits,* pp. 17-19.

Beitz, C. R. (1989). *Political equality: An essay in democratic theory*. Princeton, NJ: Princeton University Press.

Benello, C. G. (1988, Winter). Growing a local economy: Self-financing and democratic development. *Changing Work,* pp. 31-36.

Bergman, E. M. (1986). Introduction: Policy realities and development potentials in local economies. In E. M. Bergman (Ed.), *Local economies in transition* (pp. 1-14). Durham, NC: Duke University Press.

Berk, G., & Swanstrom, T. (1994, September). *Expanding the agenda of regime theory*. Paper presented at the annual meeting of the American Political Science Association, New York.

Betancur, J., Bennett, D., & Wright, P. (1991). Effective strategies for community economic development. In P. Nyden & W. Wiewel (Eds.), *Challenging uneven development: An urban agenda for the 1990s* (pp. 198-223). New Brunswick, NJ: Rutgers University Press.

Bingham, R., & Blair, J. (Eds.). (1984). *Urban economic development*. Beverly Hills, CA: Sage.

Birch, D. (1979). *The job generation process*. Cambridge, MA: Program on Neighborhood and Regional Change.

Birch, D. (1987). *Job creation in America*. New York: Free Press.

Blair, J. P., & Wechsler, B. (1984). A tale of two cities: A case of urban competition for jobs. In R. Bingham & J. Blair (Eds.), *Urban economic development* (pp. 269-282). Beverly Hills, CA: Sage.

Blakely, E. J. (1989). *Local economic development: Theory and practice*. Newbury Park, CA: Sage.

Block, F. (1987). *Revising state theory*. Philadelphia: Temple University Press.

Bluestone, B., & Harrison, B. (1982). *The deindustrialization of America*. New York: Basic Books.

Bookchin, M. (1982). *The ecology of freedom*. Palo Alto, CA: Cheshire.

Bookchin, M. (1987). *The rise of urbanization and the decline of citizenship*. San Francisco: Sierra Club Books.

Bowman, A. O'M. (1987a). *Tools and targets*. Washington, DC: National League of Cities.

Bowman, A. O'M. (1987b). *The visible hand*. Washington, DC: National League of Cities.

Bowman, A. O'M. (1988). Competition for economic development among Southeastern cities. *Urban Affairs Quarterly, 23,* 511-527.

Boyte, H. (1980). *The backyard revolution.* Philadelphia: Temple University Press.

Bradford, C., & Cincotta, G. (1992). The legacy, the promise, and the unfinished agenda. In G. D. Squires (Ed.), *From redlining to reinvestment: Community responses to urban disinvestment* (pp. 228-286). Philadelphia: Temple University Press.

Bradford, C., Cincotta, G., Finney, L., Hallett, S., & McKnight, J. (1981). Structural disinvestment: A problem in search of a policy. In R. Friedman & W. Schweke (Eds.), *Expanding the capacity to produce* (pp. 125-145). Washington, DC: Corporation for Enterprise Development.

Brandl, J., & Brooks, R. (1982). Public-private cooperation for urban revitalization: The Minneapolis and Saint Paul experience. In R. S. Fosler & R. A. Berger (Eds.), *Public-private partnerships in American cities* (pp. 163-199). Lexington, MA: D. C. Heath.

Brimmer, A. F., & Terrell, H. S. (1969). *The economic potential of black capitalism.* Paper presented at the 82nd annual meeting of the American Economic Association,

Brintnall, M. (1989). Future directions for federal urban policy. *Journal of Urban Affairs, 11,* pp. 1-19.

Brown, C., Hamilton, J., & Medoff, J. (1990). *Employers large and small.* Cambridge, MA: Harvard University Press.

Brown, J. (1993). Third sector enterprises in the United Kingdom and Australia. In D. Fasenfest (Ed.), *Community Economic Development* (pp. 205-221). London: Macmillan.

Browning, E. K., & Browning, J. M. (1983). *Public finance and the price system.* New York: Macmillan.

Bruyn, S. T. (1987). Beyond the market and the state. In S. T. Bruyn & J. Meehan (Eds.), *Beyond the market and the state: New directions in community economic development* (pp. 3-27). Philadelphia: Temple University Press.

Bruyn, S. T., & Meehan, J. (Eds.). (1987). *Beyond the market and the state: New directions in community economic development.* Philadelphia: Temple University Press.

Butler, J. S. (1991). *Entrepreneurship and self-help among black Americans.* Albany: State University of New York Press.

Carlson, V., & Wiewel, W. (1991, Fall). Strategic planning in Chicago. *Economic Development Commentary, 15,* 17-22.

Carnoy, M., & Shearer, D. (1980). *Economic democracy: The challenge of the 1980s.* New York: M. E. Sharpe.

Carr, D., & Wieffering, E. (1986, February 19). St. Paul Report Card. *Minneapolis-St. Paul CityBusiness, 22,* 1-6.

Case, J. (1989, January). The disciples of David Birch. *Inc.,* 39-44.

Ceaser, J. (1990). *Liberal democracy and political science.* Baltimore: Johns Hopkins University Press.

Chalkley, T. (1991, April 5). Collision course. *Baltimore City Paper,* pp. 10-19.

Clark, G. (1985). *Judges and the cities.* Chicago: University of Chicago Press.

Clark, K. (1992, August 16). In business for themselves. *Baltimore Sun.*

Clarke, S. E. (1993). The new localism: Local politics in a global era. In E. Goetz & S. Clarke (Eds.), *The new localism* (pp. 1-21). Newbury Park, CA: Sage.

Clarke, S. E. (1991). The changing terrain of local politics. *Urban Affairs Quarterly, 26,* 457-465.

Clarke, S. E. (1987). More autonomous policy orientations. In C. Stone & H. Sanders (Eds.), *The politics of urban development* (pp. 105-124). Lawrence: University Press of Kansas.

Clarke, S. E., & Gaile, G. A. (1989). Moving toward entrepreneurial economic development policies: Opportunities and barriers. *Policy Studies Journal, 17* 574-598.

Clarke, S. E., & Gaile, G. A. (1992). The next wave: Postfederal local economic development strategies. *Economic Development Quarterly, 6,* 187-198.

Clavel, P. (1986). *The progressive city: Planning and participation, 1969-1984.* New Brunswick, NJ: Rutgers University Press.

Clavel, P., & Kleniewski, C. (1990). Space for progressive local policy: Examples from the U.S. and U.K. In J. Logan & T. Swanstrom (Eds.), *Beyond the city limits* (199-236). Philadelphia: Temple University Press.

Clavel, P., & Wiewel, W. (1991). Introduction to P. Clavel & W. Wiewel (Eds.), *Harold Washington and the neighborhoods: Progressive city government in Chicago, 1983-1987* (pp. 1-33). New Brunswick, NJ: Rutgers University Press.

Cohen, C. J., & Dawson, M. C. (1993). Neighborhood poverty and African American politics. *American Political Science Review, 87,* 286-302.

Coleman, M. (1983). *Interest intermediation and local urban development.* Unpublished doctoral dissertation, University of Pittsburgh.

Colton, R. D., & Fisher, P. S. (1987). Public inducement of local economic development: Legal constraints on government equity. *Journal of Urban and Contemporary Law, 31,* 45-75.

Committee for Economic Development. (1982). *Public-private partnership: An opportunity for urban communities.* New York: Community for Economic Development.

Conn, D. (1992, February 16). Use of public pension funds sparks debate. *Baltimore Sun.*

Conroy, W. R. (1990). *Challenging the boundaries of reform.* Philadelphia: Temple University Press.

Conte, M., & Tannenbaum, A. (1978, July). Employee-owned companies: Is the difference measurable? U.S. Bureau of Labor Statistics. *Monthly Labor Review,* pp. 23-28.

Corcoran, E., & Wallich, P. (1991, August). The rise and fall of cities. *Scientific American,* p. 103.

Cosgriff, B. T. (1989). Small-business incubator. In D. A. Lanegran, C. Seelhammer, & A. L. Walgrave (Eds.), *The Saint Paul experiment: Initiatives of the Latimer administration* (pp. 205-214). St. Paul, MN: City of Saint Paul.

Covington, S. (1989). CDCs: Community organizations or neighborhood developers? *National Civic Review, 78,* 178-186.

Cox, K. R., & Mair, A. (1988). Locality and community in the politics of local economic development. *Annals of the Association of American Geographers, 78,* 307-325.

Cullen, B. W. (1989a). The Saint Paul energy mobilization: Pomp and circumstance. In D. A. Lanegran, C. Seelhammer, & A. L. Walgrave (Eds.), *The Saint Paul experiment: Initiatives of the Latimer administration* (pp. 5-22). St. Paul, MN: City of Saint Paul.

Cullingworth, J. B. (1987). *Urban and regional planning in Canada.* New Brunswick, NJ: Transaction Books.

Cummings, S. (1980). Collectivism: The unique legacy of immigrant economic development. In S. Cummings (Ed.), *Self-help in urban America* (pp. 5-29). Port Washington, NY: Kennikat.

Cummings, S. (Ed.). (1988). *Business elites and urban development.* Albany: State University of New York Press.

Cummings, S., Koebel, C. T., & Whitt, J. A. (1988). Public-private partnerships and public enterprise. *Urban Resources, 5,* 35-36, 47-48.

Cummings, S., Koebel, C. T., & Whitt, J. A. (1989). Redevelopment in downtown Louis-ville: Public investments, private profits, and shared risks. In G. D. Squires (Ed.), *Unequal partnerships: The political economy of urban redevelopment in postwar America* (pp. 202-221). New Brunswick, NJ: Rutgers University Press.

Cunningham, J. V. (1983). Power, participation, and local government: The communal struggle for parity. *Journal of Urban Affairs, 5*, 257-266.

Curtis, K. A., Bartelt, D. W., & Levin-Waldman, O. (1985). *Economic revitalization in the city: A sourcebook.* Philadelphia: Temple University Institute for Policy Studies.

Dahl, R. A. (1961). *Who governs?* New Haven, CT: Yale University Press.

Dahl, R. A. (1967). The city in the future of democracy. *American Political Science Review, 61*, 953-970.

Dahl, R. A. (1982). *Dilemmas of pluralist democracy.* New Haven, CT: Yale University Press.

Dahl, R. A. (1985). *A preface to economic democracy.* Berkeley: University of California Press.

Dana, R. (1984, February 9). *Homegrown economy* [Photocopy of internal memorandum to James Bellus]. City of Saint Paul, Department of Planning and Economic Development.

Daniels, B., Barbe, N., & Seigel, B. (1981). The experience and potential of community-based development. In R. Friedman & W. Schweke (Eds.), *Expanding the capacity to produce* (pp. 176-185). Washington, DC: Corporation for Enterprise Development.

Davis, J. E. (1983). Reallocation equity: A land trust model of land reform. In C. C. Geisler & F. J. Popper (Eds.), *Land reform, American style* (pp. 209-232). Totowa, NJ: Rowan & Allanheld.

DeLeon, R. E. (1992a). *Left Coast city: Progressive politics in San Francisco, 1975-1991.* Lawrence: University Press of Kansas.

DeLeon, R. E. (1992b). The urban antiregime: Progressive politics in San Francisco. *Urban Affairs Quarterly, 27*, 555-579.

Denning, M. (n.d.). *The public ownership of productive resources: An economic analysis of government enterprise.* Unpublished manuscript, University of Washington.

DiGaetano, A. (1989). Urban political regime formation. *Journal of Urban Affairs, 11*, 261-281.

Domhoff, W. G. (1978). *Who really rules? New Haven and community power reexamined.* Santa Monica, CA: Goodyear.

Dreier, P. (1989). Economic growth and economic justice in Boston. In G. D. Squires (Ed.), *Unequal partnerships: The political economy of urban redevelopment in postwar America* (pp. 35-58). New Brunswick, NJ: Rutgers University Press.

Dryzek, J. S. (1987). *Rational ecology: Environment and political economy.* Oxford, UK: Basil Blackwell.

Duncan, W. (1986, Spring). An economic development strategy. *Social Policy*, pp. 17-24.

Eisinger, P. K. (1983). Municipal residency requirements and the local economy. *Social Science Quarterly, 64*, 85-96.

Eisinger, P. K. (1988). *The rise of the entrepreneurial state.* Madison: University of Wisconsin Press.

Eisinger, P. K. (1991). The state of state venture capitalism. *Economic Development Quarterly, 5*, 64-76.

Eisinger, P. K. (1993). State venture capitalism, state politics, and the world of high-risk investment. *Economic Development Quarterly, 7*, 131-139.

Elkin, S. L. (1985). Twentieth century urban regimes. *Journal of Urban Affairs, 7*, 11-28.

Elkin, S. L. (1987). *City and regime in the American republic.* Chicago: University of Chicago Press.

Elkin, S. L. (1990, August 30-September 2). *Business-state relations in the commercial republic.* Paper presented at the annual meeting of the American Political Science Association, San Francisco.

Elkin, S. L. (1993). Constitutionalism's successor. In S. L. Elkin & K. E. Soltan (Eds.), *A new constitutionalism: Designing political institutions for a good society* (pp. 117-143). Chicago: University of Chicago Press.

Elkin, S. L., & Soltan, K. E. (Eds.). (1993). *A new constitutionalism: Designing political institutions for a good society.* Chicago: University of Chicago Press.

Ellerman, D. P. (1987). What is a worker cooperative? In S. T. Bruyn & J. Meehan (Eds.), *Beyond the market and the state: New directions in community economic development* (pp. 234-244). Philadelphia: Temple University Press.

Ellickson, R. (1982). Cities and homeowners associations. *University of Pennsylvania Law Review, 130,* 1519-1580.

Euchner, C. C. (1992). *Playing the field: Why sports teams move and cities fight to keep them.* Baltimore: Johns Hopkins University Press.

Fainstein, N. (1993). Underclass: Over class, race, and inequality. *Urban Affairs Quarterly, 29,* 340-347.

Fainstein, S. S. (1990). The changing world economy and urban restructuring. In D. Judd & M. Parkinson (Eds.), *Leadership and urban regeneration* (pp. 31-47). Newbury Park, CA: Sage.

Fainstein, S. S., & Fainstein, N. (1995). A proposal for urban policy in the 1990s. *Urban Affairs Review, 30,* 630-634.

Fainstein, S., Fainstein, N. I., Hill, R. C., Judd, D., & Smith, M. P. (1983). *Restructuring the city.* New York: Longman.

Fallows, J. (1994). *Looking at the sun.* New York: Pantheon.

Faux, G. T. (1971). *CDCs: New hope for the inner-city.* New York: Twentieth Century Fund.

Feagin, J. R. (1988). Tallying the social costs of urban growth under capitalism. In S. Cummings (Ed.), *Business elites and urban development.* Albany: State University of New York Press.

Ferlauto, R., Stumberg, R., & Sampson, R. (1992). *State policy support for targeted pension fund investment.* Washington, DC: Center for Policy Alternatives.

Ferman, B. (1992). *Challenging the growth machine: Urban politics in transition.* Unpublished manuscript.

Ferman, B. (1996). *Challenging the growth machine.* Lawrence: University Press of Kansas.

Fiedler, T. (1990, August). The lost city. *Corporate Report Minnesota,* pp. 33-43.

Fisher, L. M. (1987, April 4). Cities turn into entrepreneurs. *New York Times,* p. 33.

Fisher, P. S. (1988). State venture capital funds as an economic development strategy. *Journal of the American Planning Association, 54,* 166-177.

Fishkin, J. (1983). *Justice, equal opportunity, and the family.* New Haven, CT: Yale University Press.

Fitzgerald, J., & Simmons, L. (1991). From consumption to production: Labor participation in grass-roots movements in Pittsburgh and Hartford. *Urban Affairs Quarterly, 26,* 512-531.

Fletcher, M. A. (1992, January 28). Residency rule considered for city employment. *Baltimore Sun.*

Fosler, R. S. (1988, April). The future economic role of local governments. *Public Management, 70,* 3-10.

Fosler, R. S., & Berger, R. A. (Eds.). (1982). *Public-private partnerships in American cities.* Lexington, MA: D. C. Heath.

Fosler, R. S., & Ehsani, K. (1992). *Cooperate to compete*. Baltimore: Johns Hopkins University Institute for Policy Studies.

Frank, P. H. (1991, September 5). Study says rates in city can be cut. *Baltimore Sun*.

Frieden, B. J. (1990). Deal making goes public: Learning from Columbus Center. In T. Lassar (Ed.), *City deal making* (pp. 45-55). Washington, DC: Urban Land Institute.

Frieden, B. J., & Sagalyn, L. B. (1989). *Downtown, Inc.: How America rebuilds cities*. Boston: MIT Press.

Friedland, R. (1983). *Power and crisis in the city*. New York: Macmillan.

Friedland, R., & Palmer, D. (1984). Park Place and Main Street: Business and the urban power structure. *Annual Review of Sociology, 10*, 393-416.

Friedman, R. (1986, November). Entrepreneurial renewal in the industrial city. *Annals of American Academy of Political and Social Science*, pp. 35-44.

Friedman, R., & Schweke, W. (Eds.). (1981). *Expanding the capacity to produce*. Washington, DC: Corporation for Enterprise Development.

Friedmann, J. (1982). Urban communes, self-management, and the reconstruction of the local state. *Journal of Planning Education and Research, 2*, 37-53.

Frug, G. E. (1980). The city as a legal concept. *Harvard Law Review, 93*, 1059-1154.

Frug, G. E. (1982). Cities and homeowners associations: A reply. *University of Pennsylvania Law Review, 130*, 1589-1601.

Frug, G. E. (1984). Property and power. *American Bar Foundation Research Journal, 3*, 673-691.

Frug, G. E. (1987). Empowering cities in a federal system. *Urban Lawyer, 19*, 553-568.

Frug, G. E. (1988). *Local government law*. St. Paul, MN: West.

Fuerbringer, J. (1992, September 22). New York City to sell "mini-bonds." *New York Times*.

Fusfeld, D. R., & Bates, T. (1984). *The political economy of the urban ghetto*. Carbondale: Southern Illinois University Press.

Gangrade, K. D. (1989). Economic and social development: A Gandhian perspective. In R. P. Misra (Ed.), *The Gandhian model of development and world peace* (pp. 53-74). New Delhi, India: Concept.

Gans, H. J. (1990). Deconstructing the underclass: The term's danger as a planning concept. *Journal of the American Planning Association, 56*, 271-277.

Ganz, A. (1986). Where has the urban crisis gone? In M. Gottdiener (Ed.), *Cities in stress* (pp. 39-58). Beverly Hills, CA: Sage.

Garb, M. (1992, August 5). Saving the inner cities. *In These Times, 18*, pp. 12-13.

Garber, J. A. (1989). *Land, law, and the political economy of the American city*. Unpublished doctoral dissertation, University of Maryland—College Park.

Garber, J. A. (1990). Law and the possibilities for a just political economy. *Journal of Urban Affairs, 12*, 1-15.

Garber, J. A., & Imbroscio, D. L. (1996). The "myth of the North American city" reconsidered: Local constitutional regimes in Canada and the United States. *Urban Affairs Review, 31*, 595-624.

George, H. (1879/1955). *Progress and poverty*. New York: Robert Schalkenbach Foundation.

Giloth, R. P. (1988). Community economic development: Strategies and practices of the 1980s. *Economic Development Quarterly, 2*, 343-350.

Giloth, R. P. (1991). Making policy with communities: Research and development in the Department of Economic Development. In P. Clavel & W. Wiewel (Eds.), *Harold Washington and the neighborhoods: Progressive city government in Chicago, 1983-1987* (pp. 100-120). New Brunswick, NJ: Rutgers University Press.

Giloth, R. P., & Mier, R. (1989). Spatial change and social justice: Alternative economic development in Chicago. In R. A. Beauregard (Ed.), *Economic restructuring and political response* (pp. 181-208). Newbury Park, CA: Sage.

Goetz, E. G. (1990). Type II policy and mandated benefits in economic development. *Urban Affairs Quarterly, 25,* 170-190.

Goetz, E. G. (1992). Local government support for nonprofit housing: A survey of U.S. cities. *Urban Affairs Quarterly, 27,* 420-435.

Goetz, E. G. (1994). Expanding possibilities in local development policy. *Political Research Quarterly, 47,* 85-109.

Goldsmith, W. W., & Blakely, E. J. (1992). *Separate societies: Poverty and inequality in U.S. cities.* Philadelphia: Temple University Press.

Goodman, R. (1972). *After the planners.* New York: Simon & Schuster.

Goodman, R. (1979). *The last entrepreneurs.* New York: Simon & Schuster.

Goodsell, C. (1985). *The case for bureaucracy.* Chatham, NJ: Chatham House.

Gottdiener, M. (1987). *The decline of local politics.* Newbury Park, CA: Sage.

Greenberg, E. S. (1986). *Workplace democracy.* Ithaca, NY: Cornell University Press.

Greene, K. V., & Moulton, G. D. (1986). Municipal residency requirement statutes: An economic analysis. *Research in Law and Economics, 9,* 185-204.

Greer, J. (1987). The political economy of the local state. *Politics and Society, 15,* 513-538.

Gunn, C. (1992). Plywood co-operatives in the United States. *Economic and Industrial Democracy, 13,* 525-534.

Gunn, C., & Gunn, H. D. (1991). *Reclaiming capital: Democratic initiatives and community development.* Ithaca, NY: Cornell University Press.

Gutmann, A. (1980). *Liberal equality.* Cambridge, UK: Cambridge University Press.

Gutmann, A. (1983). How liberal is democracy? In D. MacLean & C. Mills (Eds.), *Liberalism reconsidered.* Totowa, NJ: Rowan & Allanheld.

Hager, C. M. (1980). Residency requirements for municipal employees: Important incentives in today's urban crisis. *Urban Law Annual, 18,* 197-222.

Haider, D. (1992). Place wars: New realities of the 1990s. *Economic Development Quarterly, 6,* 127-134.

Hambleton, R. (1990). Future directions for urban government in Britain and America. *Journal of Urban Affairs, 12,* 75-94.

Hammond, J. (1987). Consumer cooperatives. In S. T. Bruyn & J. Meehan (Eds.), *Beyond the market and the state: New directions in community economic development* (pp. 97-112). Philadelphia: Temple University Press.

Harrison, B. (1974). *Urban economic development: Suburbanization, minority opportunity, and the condition of the central city.* Washington, DC: Urban Institute.

Hartog, H. (1983). *Public property and private power: The corporation of the city of New York in American law.* Chapel Hill: University of North Carolina Press.

Harvey, D. (1973). *Social justice and the city.* Baltimore: Johns Hopkins University Press.

Harvey, D. (1991). Flexibility: Threat or opportunity. *Socialist Review, 21,* 65-77.

Hasell, S. (1994, January 9). From a Bronx loan plan, frustrations and trinkets. *New York Times*

Hatch, C. R. (1991). The power of manufacturing networks. In R. S. Fosler (Ed.), *Local economic development* (pp. 33-40). Washington, DC: International City Management Association.

Hatch, C. R. (1986, April). Reviving local manufacturing, Italian style. *City limits,* pp. 16-19.

Hendricks, H. (1989). *Saint Paul small-business incubator report: History, program, intent, and objectives.* City of Saint Paul: Department of Planning and Economic Development, Neighborhood Development Division.

Heilbroner, R. (1988). *Behind the veil of economics*. New York: Norton.

Henig, J. R. (1982). *Neighborhood mobilization*. New Brunswick, NJ: Rutgers University Press.

Henig, J. R. (1992). Defining city limits. *Urban Affairs Quarterly, 27*, 375-395.

Herrero, T. (1991). Housing linkage: Will it play a role in the 1990s? *Journal of Urban Affairs, 13*, 1-19.

Hill, R. C. (1983). Crisis in the Motor City. In S. Fainstein, N. I. Fainstein, R. C. Hill, D. Judd, D., & M. P. Smith (Eds.), *Restructuring the city* (pp. 80-125). New York: Longman.

Hochschild, J. L. (1991). The politics of the estranged poor. *Ethics, 101*, 560-578.

Holusha, J. (1995, December 4). Cities redeveloping old industrial sites with EPA's aid. *New York Times*.

Homewood-Brushton Revitalization and Development Corporation. (1990). *Grant request for a site feasibility study*. Homewood-Brushton Revitalization and Development Corporation, Pittsburgh, PA.

Homewood-Brushton Revitalization and Development Corporation. (n.d.). *The history of the Homewood-Brushton community*. Homewood-Brushton Revitalization and Development Corporation, Pittsburgh.

Horan, C. (1991). Beyond governing coalitions: Analyzing urban regimes in the 1990s. *Journal of Urban Affairs, 13*, 119-135.

Hornack, J. S., & Lynd, S. (1987). The Steel Valley Authority. *Review of Law and Social Change, 15*, 113-125.

Howard, C., Lipsky, M., & Marshall, D. R. (1994). Citizen participation in urban politics. In G. E. Peterson (Ed.), *Big-city politics, governance, and fiscal constraints* (pp. 153-199). Washington, DC: Urban Institute.

Humphrey, C. R., Erickson, R., & Ottensmeyer, E. J. (1989). Industrial development organizations and the local dependence hypothesis. *Policy Studies Journal, 17*, 624-642.

Hunter, F. (1980). *Community power succession: Atlanta's policy-makers revisited*. Chapel Hill: University of North Carolina Press.

Imbroscio, D. L. (1993). Overcoming the economic dependence of urban America. *Journal of Urban Affairs, 15* 173-190.

Imbroscio, D. L. (1995a). An alternative approach to urban economic development: Exploring the dimensions and prospects of a "self-reliance" strategy. *Urban Affairs Review, 30*, 840-867.

Imbroscio, D. L. (1995b). Nontraditional public enterprise as local economic development policy. *Policy Studies Journal, 23*, 218-230.

Institute for Fiduciary Education. (1989). *Economically targeted investments: A reference for public pension funds*. A report sponsored by the Ford Foundation, Sacramento, CA.

Jacobs, J. (1961). *The death and life of great American cities*. New York: Random House.

Jacobs, J. (1969). *The economy of cities*. New York: Random House.

Jacobs, J. (1984). *Cities and the wealth of nations: Principles of economic life*. New York: Random House.

Jezierski, L. (1988). Political limits to development in two declining cities: Cleveland and Pittsburgh. In M. Wallace & J. Rothschild (Eds.), *Research in politics and society* (Vol. 3, pp. 173-189). Greenwich, CT: JAI.

Jezierski, L. (1990). Neighborhoods and public-private partnerships in Pittsburgh. *Urban Affairs Quarterly, 26*, 217-249.

Johnson, A. T. (1991). Local government, minor league baseball, and economic development strategies. *Economic Development Quarterly, 5*, 313-324.

Jonas, A. E. G. (1993). Urban theory: Reworking the division of labor. *Urban Geography,* *14,* 397-407.

Jones, B., & Bachelor, L. (1984). Local policy discretion and the corporate surplus. In R. Bingham & J. Blair (Eds.), *Urban economic development* (pp. 245-267). Beverly Hills, CA: Sage.

Jones, B., & Bachelor, L. (1986). *The sustaining hand.* Lawrence: University Press of Kansas.

Judd, D. R. (1984). *The politics of American cities.* Boston: Little, Brown.

Judd, D. R., & Ready, R. L. (1986). Entrepreneurial cities and the new policies of economic development. In G. Peterson & C. W. Lewis (Eds.), *Reagan and the cities* (pp. 209-247). Washington, DC: Urban Institute Press.

Judd, D. R., & Swanstrom, T. (1988). Business and cities: The enduring tension. *Urban Resources, 5,* pp. 3-8, 44-46.

Kantor, P. (1988). *The dependent city.* Glenview, IL: Scott, Foresman/Little, Brown.

Kantor, P., & Savitch, H. V. (1993). Can politicians bargain with business? *Urban Affairs Quarterly, 29,* 230-255.

Kasarda, J. D. (1985). Urban change and minority opportunities. In P. Peterson (Ed.), *The new urban reality* (pp. 33-67). Washington, DC: Brookings Institution.

Kasarda, J. D. (1989). Urban industrial transition and the underclass. In W. J. Wilson (Ed.), *The ghetto underclass: Social science perspectives* (pp. 26-47). Newbury Park, CA: Sage.

Katznelson, I. (1981). *City trenches: Urban politics and the patterning of class in the United States.* Chicago: University of Chicago Press.

Kaus, M. (1992). *The end of equality.* New York: Basic Books.

Keating, D., Krumholz, N., & Metzger, J. (1989). Cleveland: Post-populist public-private partnerships. In G. D. Squires (Ed.), *Unequal partnerships: The political economy of urban redevelopment in postwar America* (pp. 121-141). New Brunswick, NJ: Rutgers University Press.

Keating, M. (1991). *Comparative urban politics.* Hants, UK: Edward Elgar.

Kelly, R. M. (1977). *Community control of economic development.* New York: Praeger.

Kershaw, A. (1990, January 9). The bartered bride. *The Bulletin,* pp. 68-69.

Kieschnick, M. (1981). The role of equity capital in urban economic development. In R. Friedman & W. Schweke (Eds.), *Expanding the capacity to produce* (pp. 374-386). Washington, DC: Corporation for Enterprise Development.

Kotler, M. (1969). *Neighborhood government: The local foundation of political life.* Indianapolis, IN: Bobbs-Merrill.

Kowinski, W. S. (1993, October). The tale of city pride: People who wouldn't give up. *Smithsonian,* pp. 118-130.

Krasner, S. D. (1978). *Defending national interest: Raw materials investment and U.S. foreign policy.* Princeton, NJ: Princeton University Press.

Krumholz, N. (1991). Equity and local economic development. *Economic Development Quarterly, 5,* 291-300.

Landsman Community Services Ltd. (1989). *Local currencies.* Courtenay, BC: Landsman Community Services Ltd.

Lane, P. (1988). Community-based economic development: Our Trojan horse. *Studies in Political Economy, 26,* 177-191.

Lanegran, D. A., Seelhammer, C., & Walgrave, A. L. (Eds.). (1989). *The Saint Paul experiment: Initiatives of the Latimer administration.* St. Paul, MN: City of Saint Paul.

Lassar, T. (1990). Introduction. In T. Lassar (Ed.), *City deal making* (pp. 1-7). Washington, DC: Urban Land Institute.

Latimer, G. (1989). Introduction. In D. A. Lanegran, C. Seelhammer, & A. L. Walgrave (Eds.), *The Saint Paul experiment: Initiatives of the Latimer administration* (pp. xix-xxii). St. Paul, MN: City of Saint Paul.

Leatherwood, T. (1983). Pension fund investment. In *America's cities and counties: A citizen's agenda* (pp. 21-26). Washington, DC: Conference on Alternative State and Local Policies.

Leitner, H., & Garner, M. (1993). The limits of local initiatives: A reassessment of urban entrepreneurialism for urban development. *Urban Geography, 14*, 57-77.

Lemann, N. (1991). *The promise land.* New York: Vintage.

Levine, M. V. (1987). Downtown redevelopment as an urban growth strategy: A critical appraisal of the Baltimore renaissance. *Journal of Urban Affairs, 9*, 103-123.

Levine, M. V. (1988). Economic development in states and cities: Toward democratic and strategic planning in state and local government. In M. V. Levine, C. MacLennan, J. J. Kushma, & C. Noble, *The state and democracy* (pp. 111-146). New York: Routledge.

Levine, M. V. (1989). The politics of partnership: Urban redevelopment since 1945. In G. D. Squires (Ed.), *Unequal partnerships: The political economy of urban redevelopment in postwar America* (pp. 12-34). New Brunswick, NJ: Rutgers University Press.

Levy, J. M. (1992). The U.S. experience with local economic development. *Environment and Planning C: Government and policy, 10*, 51-60.

Lindblom, C. E. (1965). *The intelligence of democracy.* New York: Free Press.

Lindblom, C. E. (1977). *Politics and markets.* New York: Basic Books.

Lindblom, C. E. (1982). The market as prison. *Journal of Politics, 44*, 324-336.

Linsalata, P., & Novak, T. (1992, January 26). Pittsburgh is finding a way to revitalize its decaying neighborhoods. *Chicago Tribune.*

Lipietz, A. (1986). New tendencies in the international division of labor: Regions of accumulation and modes of regulation. In A. J. Scott & M. Storper (Eds.), *Production, work, and territory: The geographical anatomy of industrial capitalism* (pp. 16-40). Boston: Allen & Unwin.

Litvak, L., & Daniels, B. (1979). *Innovations in development finance.* Washington, DC: Council of State Planning Agencies.

Logan, J. R., & Molotch, H. (1987). *Urban fortunes.* Berkeley: University of California Press.

Logan, J. R., & Swanstrom, T. (1990). Urban restructuring: A critical view. In J. R. Logan & T. Swanstrom (Eds.), *Beyond the city limits* (pp. 3-24). Philadelphia: Temple University Press.

Long, N. E. (1962). *The polity.* Chicago: Rand McNally.

Long, N. E. (1972). *The unwalled city.* New York: Basic Books.

Long, N. E. (1983, March). *Can the contemporary city be a significant polity?* Paper presented at the annual meeting of the Urban Affairs Association, Flint, MI.

Long, N. E. (1986). The city as a political community. *Journal of Community Psychology, 14*, 72-80.

Long, N. E. (1987). Labor intensive and capital intensive urban economic development. *Economic Development Quarterly, 1*, 196-202.

Lubove, R. (1969). *Twentieth-century Pittsburgh.* New York: John Wiley.

Lueck, T. J. (1992, August 11). Dinkins plan for loans to businesses. *New York Times.*

Lurcott, R. H., & Downing, J. A. (1987). A public-private support system for community-based organizations in Pittsburgh. *Journal of the American Planning Association, 53*, 459-468.

Luria, D., & Russell, J. (1982). Rebuilding Detroit: A rational reindustrialization strategy. *Socialist Review, 12,* 163-183.

Lustig, R. J. (1985). The politics of shutdown: Community, property, corporatism. *Journal of Economic Issues, 19,* 123-151.

Lynd, S. (1981, July). Reindustrialization: Brownfield or greenfield. *Democracy, 1,* 22-36.

Lynd, S. (1987a). The genesis of the idea of a community right to industrial property in Youngstown and Pittsburgh, 1977-1987. *Journal of American History, 74,* 926-58.

Lynd, S. (1987b). Towards a not-for-profit economy: Public development authorities for acquisition and use of industrial property. *Harvard Civil Rights—Civil Liberties Law Review, 22,* 13-41.

Lynd, S. (1989, Spring). The genesis of the idea of a community right to industrial property in Youngstown and Pittsburgh, 1977-1987. *Changing Work,* 14-19.

MacDonald, D. (1985, September 2). Incubator fever. *New England Business,* 62-72.

Macpherson, C. B. (1977). *The life and times of liberal democracy.* Oxford, UK: Oxford University Press.

Magnusson, W. (1989, August 31-September 3). *Radical municipalities in North America.* Paper presented at the annual meeting of the American Political Science Association, Atlanta, GA.

Malizia, E. E. (1985). *Local economic development: A guide to practice.* New York: Praeger.

Mandelker, D. R., Netsch, D. C., Salsich, Jr., P. W., & Wagner, J. W. (1990). *State and local government in a federal system* (3rd ed.). Charlottesville, VA: Michie.

Mansbridge, J. J. (1977). Acceptable inequalities. *British Journal of Political Science, 7,* 321-336.

Mansbridge, J. J. (1980). *Beyond adversary democracy.* Chicago: University of Chicago Press.

Mansbridge, J. J. (Ed.). (1990). *Beyond self-interest.* Chicago: University of Chicago.

Margolis, R. (1983, May-June). Reaganomics redux: A municipal report. *Working Papers for a New Society, 10,* 41-47.

Markusen, A. R. (1988). Planning for industrial decline: Lessons from steel communities. *Journal of Planning Education and Research, 7,* 173-184.

Marquez, B. (1993). Mexican-American community development corporation and the limits of directed capitalism, *Economic Development Quarterly 7,* 287-295.

Marsh, T. R., & McAllister, D. E. (1981). ESOPs tables: A survey of companies with employee stock ownership plans. *Journal of Corporate Law, 6,* 551-623.

Martin, D. G. (1989). District heating. In D. A. Lanegran, C. Seelhammer, & A. L. Walgrave (Eds.), *The Saint Paul experiment: Initiatives of the Latimer administration* (pp. 259-267). St. Paul, MN: City of Saint Paul.

Mayer, M. (1988, June). *The changing conditions for local politics in the transition to post-Fordism.* Paper presented at the International Conference on Regulation Theory, Barcelona, Spain.

Mayer, M. (1991). Politics in the post-Fordist city. *Socialist Review, 21,* 105-123.

Mayer, N. (1984). *Neighborhood organizations and community development: Making revitalization work.* Washington, DC: Urban Institute Press.

McArthur, A. A. (1993). Community business and urban regeneration. *Urban Studies, 30,* 849-873.

McCarthy, D. J. (1990). *Local government law* (3rd ed.). St. Paul, MN: West.

McCarroll, T. (1992, January 6). Entrepreneurs. *Time,* pp. 62-65.

McClelland, D. C. (1961). *The achieving society.* Princeton, NJ: Van Nostrand.

McCormick, J. (1989, February 6). America's hot cities: The Twin Cities' better half. *Newsweek,* p. 43.

McGuire, M. J. (1992, August 30). Making money the old-fashioned way. *Albany Sunday Times Union*.

Meehan, J. (1987). Working toward local self-reliance. In S. T. Bruyn & J. Meehan (Eds.), *Beyond the market and the state: New directions in community economic development* (pp. 131-151). Philadelphia: Temple University Press.

Metzger, J. T. (1992). The Community Reinvestment Act and neighborhood revitalization in Pittsburgh. In G. D. Squires (Ed.), *From redlining to reinvestment: Community responses to urban disinvestment* (pp. 73-108). Philadelphia: Temple University Press.

Meyer, P. B. (1991). Local economic development: What is proposed, what is done, and what difference does it make? *Policy Studies Review, 10*, 172-180.

Mier, R. (1993). *Social justice and local development policy*. Newbury Park, CA: Sage.

Mollenkopf, J. H. (1983). *The contested city*. Princeton, NJ: Princeton University Press.

Mollenkopf, J. H. (1989). Who (or what) runs cities, and how? *Sociological Forum, 4*, 119-137.

Molotch, H. (1976). The city as a growth machine. *American Journal of Sociology, 82*, 309-332.

Monti, D. J. (1990). *Race, redevelopment, and the new company town*. Albany: State University of New York Press.

Morris, D. (1982a). *The new city-states*. Washington, DC: Institute for Local Self-Reliance.

Morris, D. (1982b). *Self-reliant cities*. San Francisco: Sierra Club Books.

Morris, D. (1985). *The homegrown economy: A prescription for Saint Paul's future*. Washington, DC: Institute for Local Self-Reliance.

Morris, D. (1986, February 3). *Final report: Self-reliant city effort*. City of Saint Paul, St. Paul, MN.

Morris, D. (1991, November/December). You can *fight* city hall. *Utne Reader*, pp. 89-92.

Morris, D., & Hess, K. (1975). *Neighborhood power*. Boston: Beacon.

Muzzio, D., & Bailey, R. W. (1986). Economic development, housing and zoning: A tale of two cities. *Journal of Urban Affairs, 8*, 1-18.

Nakano, D., & Williamson, T. (1993). *Uniting economics politics, democracy, and community: Community-based building blocks of a democratic economy*. Unpublished manuscript, National Center for Policy Alternatives, Washington, DC.

Nathan, R. P. (1989). Institutional change and the challenge of the underclass. In W. J. Wilson (Ed.), *The ghetto underclass: Social science perspectives* (pp. 170-181). Newbury Park, CA: Sage.

National Congress for Community Economic Development. (1989). *Against all odds: The achievements of community-based development organizations*. Washington, DC: National Congress for Community Economic Development.

Newman, H. (1993, April). *Community development corporations and neighborhood revitalization in Atlanta*. Paper presented at the annual meeting of the Urban Affairs Association, Indianapolis, IN.

Nickel, D. R. (1995). The progressive city? *Urban Affairs Review, 30*, 355-377.

Norman, R. T. (1989). Empowering cities and citizens. In R. P. Misra (Ed.), *The Gandhian model of development and world peace* (pp. 223-240). New Delhi, India: Concept.

North Side Civic Development Council. (1991). *NSCDC activity summary*. North Side Civic Development Council, Pittsburgh, PA.

Nove, A. (1983). *The economics of feasible socialism*. London: George Allen & Unwin.

Noyelle, T. (1986, November). Economic transformation. *Annals of the American Academy of Political and Social Science*, 9-17.

Oakland Planning and Development Council. (1990). *Ten-year report: 1980-1990.* Pittsburgh, PA: Oakland Planning and Development Council.

Office of Mayor, Saint Paul. (1983). *Saint Paul's homegrown economy project: A new economic policy and program for a self-reliant city.* St. Paul, MN: City of Saint Paul.

Olson, D. G. (1987). Employee ownership: An economic development tool for anchoring capital in local communities. *Review of Law and Social Change, 15,* 239-267.

Olson, M. (1982). *The rise and decline of nations.* New Haven, CT: Yale University Press.

Orr, M. (1992). Urban regimes and human capital policies: A study of Baltimore. *Journal of Urban Affairs, 14,* 173-187.

Osborne, D. (1985, September). The most entrepreneurial city in America. *Inc.,* pp. 54-60.

Osborne, D. (1988). *Laboratories of democracy.* Cambridge, MA: Harvard Business School Press.

Osborne, D., & Gaebler, T. (1992). *Reinventing government.* New York: Addison-Wesley.

Oxnevad, K. (1991, March). Standard & Poor's cuts St. Paul Port Authority's $332 million of debt to BBB from BBB-plus. *The Bond Buyer, 13,* p. 4.

Papatola, D. P. (1990, October 22). Watershed years at the port authority. *Minneapolis-Saint Paul CityBusiness, 8,* 15-25.

Parzen, J. A., & Kieschnick, M. H. (1992). *Credit where it's due: Development banking for communities.* Philadelphia: Temple University Press.

Pateman, C. (1970). *Participation and democratic theory.* Cambridge, UK: Cambridge University Press.

Patton, W., & Markusen, A. (1991). The perils of overstating service sector growth potential. *Economic Development Quarterly, 5,* 197-212.

Pecorella, R. F. (1985). Resident participation as agenda setting: A study of neighborhood-based development corporations. *Journal of Urban Affairs, 7,* 13-27.

Peirce, N. R., & Steinbach, C. (1987). *Corrective capitalism: The rise of America's community development corporations.* New York: Ford Foundation.

Peretz, P. (1986). The market for incentives: Where angels fear to tread. *Policy Studies Journal, 5,* 624-633.

Perry, S. E. (1987). *Communities on the way: Rebuilding local economies in the United States and Canada.* Albany: State University of New York Press.

Persky, J., Ranney, D., & Wiewel, W. (1993). Import substitution and local economic development. *Economic Development Quarterly, 7,* 18-29.

Peterson, P. (1981). *City limits.* Chicago: Chicago University Press.

Peterson, P., & Greenstone, J. D. (1977). Racial change and citizen participation: The mobilization of low-income communities through community action. In R. H. Haveman (Ed.), *A decade of federal antipoverty programs* (pp. 241-278). New York: Academic Press.

Piore, M., & Sabel, C. (1984). *The second industrial divide.* New York: Basic Books.

Pittsburgh Partnership for Neighborhood Development. (1990). *1989-1990 Annual Report.* Pittsburgh, PA: Pittsburgh Partnership for Neighborhood Development.

Plosila, W. H., & Allen, D. (1985). Small business incubators and public policy: Implications for state and local development strategies. *Policy Studies Journal, 13,* 729-734.

Polanyi, K. (1944/1957). *The great transformation: The political and economic origins of our time.* Boston: Beacon.

Polsky, A. (1988). Jane Jacobs and the limits of urban capitalism. *Urban Resources, 5,* 9-14, 24.

Porter, M. E. (1995, May-June). The competitive advantage of the inner city. *Harvard Business Review,* pp. 55-71.

Portz, J. (1989). Plant closings: New roles for policymakers. *Economic Development Quarterly, 67*, 70-80.

Portz, J. (1990). *The politics of plant closings.* Lawrence: University Press of Kansas.

Preteceille, E. (1990). Political paradoxes of urban restructuring: Globalization of the economy and localization of politics? In J. R. Logan & T. Swanstrom (Eds.), *Beyond the city limits* (pp. 27-59). Philadelphia: Temple University Press.

Purdy, M., & Sexton, J. (1995, September 11). Facing bank service shortage, many immigrants improvise. *New York Times.*

Quarrey, M. (1986). *Employee ownership and corporate performance.* Oakland, CA: National Center for Employee Ownership.

Raven, R. F. (1991). How bankers view CRA. *Economic Development Commentary, 15*, 23-29.

Reed, A., Jr. (1988a). The black urban regime: Structural origins and constraints. In M. P. Smith (Ed.), *Power, community, and the city, comparative urban and community research* (Vol. 1, pp. 138-189). New Brunswick, NJ: Transaction.

Reed, A., Jr. (1988b, February). The liberal technocrat. *The Nation, 6*, 167-170.

Rich, M. J. (1992). UDAG, economic development, and the death and life of American cities. *Economic Development Quarterly, 6*, 150-172.

Rifkin, J., & Barber, R. (1978). *The North will rise again: Pension, politics, and power in the 1980s.* Boston: Beacon.

Riposa, G., & Andranovich, G. (1988). Economic development policy: Whose interests are being served? *Urban Resources, 5*, 25-34, 42.

Roberts, B. F. (1980). *Community development corporations and state development policy.* Washington, DC: National Congress for Community Development.

Roberts, S. (1992, June 1). Money and other excuses hinder inner-city growth. *New York Times.*

Robinson, C. J. (1989). Municipal approaches to economic development: Growth and distribution policy. *Journal of the American Planning Association, 55*, 283-294.

Robson, B. (1992, February). The neglected twin. *Mpls.st.paul, 20* 52-57, 110-111.

Rosdil, D. L. (1991). The context of radical populism in U.S. cities: A comparative analysis. *Journal of Affairs, 13*, 77-96.

Rosen, C., & Wilson, J. (1987). Employee ownership: A new strategy for economic development. *Review of Law and Social Change, 15*, 211-225.

Rosen, C., & Klein, K. (1983, August). Job creating performance of employee-owned firms, U.S. Bureau of Labor Statistics. *Monthly Labor Review*, pp. 15-19.

Rosen, D. P. (1988). *Public capital.* Washington, DC: National Center for Policy Alternatives.

Rosenthal, C. N., & Washington, D. A. (1987). Community finance in the age of Gramm-Rudman. *Review of Law and Social Change, 15*, 183-195.

Ross, S. J. (1982). Political economy for the masses: Henry George. *Democracy, 2*, 125-134.

Rubin, I. S., & Rubin, H. J. (1987). Economic development incentives: The poor (cities) pay more. *Urban Affairs Quarterly, 23*, 37-42.

Sabel, C. F. (1982). *Work and politics: The division of labor in industry.* Cambridge, UK: Cambridge University Press.

Sagalyn, L. B. (1990). Public profit sharing: Symbol or substance. In T. Lassar (Ed.), *City deal making* (pp. 139-153). Washington, DC: Urban Land Institute.

Saint Paul Port Authority. (1991). *Saint Paul Port Authority: A reference guide.* St. Paul, MN: Author.

Sale, K. (1980). *Human scale.* New York: G. P. Putnam.

Salpukas, A. (1995, October 10). The rebellion in "Pole City." *New York Times.*

Sandel, M. (1982). *Liberalism and limits of justice.* Cambridge, UK: Cambridge University Press.

Savitch, H. V., & Thomas, J. C. (1991). Conclusion: End of the millennium big city politics. In H. V. Savitch & J. C. Thomas (Eds.), *Big city politics in transition* (pp. 235-251). Beverly Hills, CA: Sage.

Sbragia, A. (1989). The Pittsburgh model of economic development: Partnership, responsiveness, and indifference. In G. D. Squires (Ed.), *Unequal partnerships: The political economy of urban redevelopment in postwar America* (pp. 103-120). New Brunswick, NJ: Rutgers University Press.

Sbragia, A. (1990). Pittsburgh's "third way": The nonprofit sector as a key to urban regeneration. In D. Judd & M. Parkinson (Eds.), *Leadership and urban regeneration* (pp. 51-68). Newbury Park, CA: Sage.

Schoettle, F. P. (1990). What public finance do state constitutions allow? In R. Bingham et al. (Eds.). *Financing economic development: An institutional response* (pp. 57-74). Newbury Park, CA: Sage.

Schramm, R. (1987). Local, regional, and national strategies. In S. T. Bruyn & J. Meehan (Eds.). *Beyond the market and the state: New directions in community economic development* (pp. 152-170). Philadelphia: Temple University Press.

Schumacher, E. F. (1975). *Small is beautiful: Economics as if people mattered.* New York: Harper & Row.

Schumpeter, J. A. (1942/1976). *Capitalism, socialism and democracy.* New York: Harper & Row.

Schweke, W. (1983). Small business. In *America's cities and counties: A citizen's agenda* (pp. 27-33). Washington, DC: Conference on Alternative State and Local Policies.

Schweke, W. (1985). Why local governments need an entrepreneurial policy. *Public Management, 67,* 3-5.

Scott, A. J. (1988). Flexible production systems and regional development: The rise of new industrial spaces in North America and Western Europe. *International Journal of Urban and Regional Research, 12*(2), 171-186.

Shapero, A. (1981). The role of entrepreneurship in economic development at the less-than-national level. In R. Friedman & W. Schweke (Eds.), *Expanding the capacity to produce* (pp. 25-35). Washington, DC: Corporation for Enterprise Development.

Shapero, A. (1985). Entrepreneurship: Key to self-renewing economies. Reprinted in E. E. Malizia, *Local economic development: A guide to practice* (pp. 209-218). New York: Praeger.

Sharp, E. (1990). *Urban politics and administration.* New York: Longman.

Shavelson, J. (1990). *A third-way, a sourcebook: Innovations in community-owned enterprise.* Washington, DC: National Center for Economic Alternatives.

Shaw, T. C. (1994, September). *Political accountability and the activism of the black poor.* Paper presented at the annual meeting of the American Political Science Association, New York.

Shearer, D. (1989). In search of equal partnerships: Prospects for progressive urban policy in the 1990s. In G. D. Squires (Ed.), *Unequal partnerships: The political economy of urban redevelopment in postwar America* (pp. 289-307). New Brunswick, NJ: Rutgers University Press.

Shefter, M. (1985). *Political crisis/fiscal crisis.* New York: Basic Books.

Shiffman, R. (1990). *Comprehensive and integrative planning for community development.* New York: New School for Social Research, Graduate School of Management and Urban Policy, Community Development Research Center.

Shlay, A. B. (1989). Financing community. *Journal of Urban Affairs, 11,* 201-223.

Sims, C. (1992, October 13). California taps its profitable ports. *New York Times.*

Smith, M. P. (1989). The uses of linked-development policies in U.S. cities. In M. Parkinson et al. (Eds.), *Regenerating the cities: The U.K. crisis and the U.S. experience* (pp. 85-99). Glenview, IL: Scott, Foresman.

Smith, M. P., & Judd, D. R. (1984). American cities: The production of ideology. In M. P. Smith (Ed.), *Cities in transformation* (pp. 173-196). Beverly Hills, CA: Sage.

Soifer, S. D. (1990). The Burlington community land trust. *Journal of Urban Affairs, 12,* 237-252.

Sonenshein, R. J. (1988). The "city limits" debate: Another perspective. *Urban Resources, 5,* 37-38.

Squires, G. D. (1989). Public-private partnerships: Who gets what and why. In G. D. Squires (Ed.), *Unequal partnerships: The political economy of urban redevelopment in postwar America* (1-11). New Brunswick, NJ: Rutgers University Press.

Squires, G. D. (1992a). Community reinvestment: An emerging social movement. In G. D. Squires (Ed.), *From redlining to reinvestment: Community responses to urban disinvestment* (1-37). Philadelphia: Temple University Press.

Squires, G. D. (Ed.). (1992b). *From redlining to reinvestment: Community responses to urban disinvestment.* Philadelphia: Temple University Press.

Squires, G. D. (1994). *Capital and communities in black and white.* Albany: State University of New York Press.

Stanback, H., & Mier, R. (1987). Economic development for whom? The Chicago model. *Review of Law and Social Change, 15,* 11-22.

Stanback, T., & Noyelle, T. (1982). *Cities in transition.* Totowa, NJ: Allanheld-Osmun.

Stevenson, R. W. (1992, March 17). Pension funds becoming a tool for growth. *New York Times.*

Stewman, S., & Tarr, J. A. (1982). Four decades of public-private partnerships in Pittsburgh. In R. S. Fosler & R. A. Berger (Eds.), *Public-private partnerships in American cities* (pp. 59-127). Lexington, MA: D. C. Heath.

Stokey, E., & Zeckhauser, R. (1978). *A primer for policy analysis.* New York: Norton.

Stone, C. N. (1976). *Economic growth and neighborhood discontent.* Chapel Hill: University of North Carolina Press.

Stone, C. N. (1980). Systemic power in community decision making. *American Political Science Review, 74,* 978-990.

Stone, C. N. (1986). Preemptive power: Floyd Hunter's "community power structure" reconsidered. *American Journal of Political Science, 32,* 82-104.

Stone, C. N. (1987). The study of the politics of urban development. In C. Stone & H. Sanders (Eds.), *The politics of urban development* (pp. 3-22). Lawrence: University Press of Kansas.

Stone, C. N. (1988, September). *Political change and regime continuity in postwar Atlanta.* Paper presented at the annual meeting of the American Political Science Association, Washington, DC.

Stone, C. N. (1989). *Regime politics: Governing Atlanta, 1946-1988.* Lawrence: University Press of Kansas.

Stone, C. N. (1991). The hedgehog, the fox, and the new urban politics. *Journal of Urban Affairs, 13,* 289-297.

Stone, C. N., Orr, M., & Imbroscio, D. (1991). The reshaping of urban leadership in U.S. cities: A regime analysis. In M. Gottdiener & C. Pickvance (Eds.), *Urban life in transition* (pp. 222-239). Newbury Park, CA: Sage.

Stone, C. N., & Sanders, H. T. (Eds.). (1987). *The politics of urban development*. Lawrence: University Press of Kansas.

Swack, M. (1987). Community finance institutions. In S. T. Bruyn & J. Meehan (Eds.). *Beyond the market and the state: New directions in community economic development* (pp. 79-97). Philadelphia: Temple University Press.

Swaine. R. K. (1993). Public policy and employee ownership. *Policy Sciences, 36*, 289-315.

Swanstrom, T. (1985). *The crisis of growth politics*. Philadelphia: Temple University Press.

Swanstrom, T. (1986). Urban populism, fiscal crisis, and the new political economy. In M. Gottdiener (Ed.), *Cities in stress* (pp. 81-110). Beverly Hills, CA: Sage.

Swanstrom, T. (1988). Semisovereign cities: The politics of urban development. *Polity, 21*, 83-110.

Swanstrom, T. (1993). Beyond economism: Urban political economy and the postmodern challenge. *Journal of Urban Affairs, 15*, 55-78.

Taub, R. P. (1988). *Community capitalism*. Boston: Harvard Business School.

Taylor, M. (1987). *The possibility of cooperation*. Cambridge, UK: Cambridge University Press.

Terry, D. (1992, February 10). On the frigid northern plains, one city finds way to new jobs in hard times. *New York Times*,

Thompson, W. R. (1965). *A preface to urban economics*. Baltimore: Johns Hopkins University Press.

Todaro, M. P. (1985). *Economic development in the Third World* (3rd ed.). New York: Longman.

Twelvetrees, A. (1989). *Organizing for neighborhood development*. Aldershot, UK: Avebury.

U.S. General Accounting Office. (1987). *ESOPs: Little evidence of effects on corporate performance* (GAO/PEMD-88-1). Washington, DC: Author.

Unger, R. M. (1987). *False necessity: Anti-necessitarian social theory in the service of radical democracy*. Part I of *Politics, a work in constructive social theory*. Cambridge, UK: Cambridge University Press.

Vaughan, R. J. (1988). Economists and economic development. *Economic Development Quarterly, 2*, 119-123.

Vaughan, R. J., & Pollard, R. (1986). Small business and economic development. In N. Walzer & D. L. Chicoine (Eds.), *Financing economic development* (pp. 122-138). New York: Praeger.

Vidal, A. C. (1992). *Rebuilding communities: A national study of urban community development corporations*. New York: New School for Social Research, Graduate School of Management and Urban Policy, Community Development Research Center.

Vogel, R. K. (1992). *Urban political economy*. Gainesville: University Press of Florida.

Vogel, R. K., & Swanson, B. (1993). The dialogical community and economic development. In D. Fasenfest (Ed.), *Community economic development* (pp. 188-204). London: Macmillan.

Walden, G. (1989, October 30). Twin cities: Talent, prosperity and growth combine in Minneapolis-St. Paul. *Advertising Age*, pp. T1-T3.

Walsh, A. H. (1978). *The public's business: The politics and practices of government corporations*. Cambridge, MA: MIT Press.

Walzer, M. (1983). *Spheres of justice: A defense of pluralism and equality*. New York: Basic Books.

Walzer, M. (1984). Liberalism and the art of separation. *Political theory, 12*, 315-330.

Warner, S. B. (1968). *The private city*. Philadelphia: University of Pennsylvania Press.

Waste, R. J. (1993). City limits, pluralism, and urban political economy. *Journal of Urban Affairs, 15*, 445-455.

Watkins, A. (1980). *The practice of urban economics*. Beverly Hills, CA: Sage.

Wayne, L. (1992, March 14). New hope in inner cities. *New York Times*.

Webber, A. E. (1989). Energy park. In D. A. Lanegran, C. Seelhammer, & A. L. Walgrave (Eds.), *The Saint Paul experiment: Initiatives of the Latimer administration* (pp. 519-539). St. Paul, MN: City of Saint Paul.

Wegner, J. W. (1990). Utopian visions: Cooperation without conflicts in public/private ventures. In T. Lassar (Ed.), *City deal making* (pp. 57-79). Washington, DC: Urban Land Institute.

Weiher, G. R. (1989). Rumors of the demise of the urban crisis are greatly exaggerated. *Journal of Urban Affairs, 11*, 225-242.

Weinberg, M. L., Allen, D. N., & Schermerhorn, J.R., Jr. (1991). Interorganizational challenges in the design and management of business incubators. *Policy Studies Review, 10*, 149-160.

Weinberg, R. (1984). The use of eminent domain to prevent an industrial plant shutdown: The next step in an expanding power. *Albany Law Review, 49*, 95-130.

Weiss, M. A., & Metzger, J. T. (1987). Technology development, neighborhood planning, and negotiated partnerships. *Journal of the American Planning Association, 53*, 469-477.

White, K., & Matthei, C. (1987). Community land trusts. In S. T. Bruyn & J. Meehan (Eds.), *Beyond the market and the state: New directions in community economic development* (pp. 41-64). Philadelphia: Temple University Press.

White, S. B., & Osterman, J. D. (1991). Is employment growth really coming from small establishments? *Economic Development Quarterly, 5*, 241-257.

Wiewel, W., & Mier, R. (1986). Enterprise activities of not-for-profit organizations. In E. M. Bergman (Ed.), *Local economies in transition* (pp. 205-225). Durham, NC: Duke University Press.

Wiewel, W., & Rieser, N. C. (1989). The limits of progressive municipal economic development: Job creation in Chicago, 1983-1987. *Community Development Journal, 24*, 111-118.

Wiewel, W., Teitz, M., & Giloth, R. (1993). The economic development of neighborhoods and localities. In R. D. Bingham & R. Mier (Eds.), *Theories of economic development* (pp. 80-99). Newbury Park, CA: Sage.

Wiewel, W., & Weintraub, J. (1990). Community development corporation as a tool for economic development finance. In R. Bingham, E. W. Hill, & S. B. White (Eds.), *Financing economic development: An institutional response* (pp. 160-175). Newbury Park, CA: Sage.

Wilson, W. J. (1987). *The truly disadvantaged*. Chicago: University of Chicago Press.

Wilson, W. J. (Ed.). (1989). *The ghetto underclass: Social science perspectives*. Newbury Park, CA: Sage.

Wolman, H. (1988). Local economic development policy: What explains the divergence between policy analysis and political behavior? *Journal of Urban Affairs, 10*, 19-28.

Yin, J. S. (1994). *The transformation of the community development corporation: A case study of politics and institutions in Cleveland, 1967-1993*. Unpublished master's thesis, Cornell University, Ithaca, NY.

Young, I. M. (1990). *Justice and the politics of difference*. Princeton, NJ: Princeton University Press.

Zdenek, R. (1987). Community development corporations. In S. T. Bruyn & J. Meehan (Eds.), *Beyond the market and the state: New directions in community economic development* (pp. 112-130). Philadelphia: Temple University Press.

Zeitlin, M. (1982, April). Democratic investment. *Democracy*, pp. 69-80.

Zipp, J. F. (1991). The quality of jobs in small business. *Economic Development Quarterly, 5*, 9-22.

Author Index

Subject Index

About the Author

David L. Imbroscio is currently an assistant professor in the department of political science at the University of Louisville. He completed his doctoral studies at the University of Maryland—College Park, earning a PhD in political science in 1993 (under the direction of Stephen L. Elkin and Clarence N. Stone). Professor Imbroscio's research explores questions intersecting the fields of political economy, normative political theory, and urban politics. His recent work investigates the viability of alternative approaches to urban economic development and the possible responses to the structural dependencies faced by cities in the post-industrial age. These studies appear in *Urban Affairs Review*, *Policy Studies Journal*, the *Journal of Urban Affairs*, as well as several edited volumes.